Minerals and Rocks

12

Editor in Chief
P. J. Wyllie, Chicago, Ill.

Editors
W. von Engelhardt, Tübingen · T. Hahn, Aachen

Robert G. Coleman

Ophiolites

Ancient Oceanic Lithosphere?

With 72 Figures

Springer-Verlag Berlin Heidelberg New York 1977

Dr. *Robert G. Coleman*
Branch of Field Geochemistry and Petrology
Geologic Division, Geological Survey
US Department of the Interior
Menlo Park, CA 94025, U.S.A.

Volumes 1 to 9 in this series appeared under the title
Minerals, Rocks and Inorganic Materials

ISBN 3-540-08276-X Springer-Verlag   Berlin Heidelberg New York
ISBN 0-387-08276-X Springer-Verlag   New York Heidelberg Berlin

Library of Congress Cataloging in Publication Data. Coleman, Robert Griffin, 1923 —. Ophiolites. (Minerals and rocks; 12). Bibliography: p. 1. Ophiolites. I. Title. QE462.06C64   552.3   77-2778.

© by Springer-Verlag Berlin · Heidelberg 1977
Printed in Germany
Typesetting, printing and binding: Brühlsche Universitätsdruckerei, Lahn-Gießen.
2132/3130-543210

Preface

I was invited to write this book as part of the Minerals, Rocks and Organic Materials Series of Springer-Verlag by Professor Peter J. Wyllie in 1974. Ophiolites have preoccupied me ever since 1948 as a graduate student and up to the present time as part of my research with the U.S. Geological Survey. During this period ophiolite, an obscure European geological term, has attained an ever-increasing importance, is now used to include all fragments of ancient oceanic lithosphere incorporated into the orogenic zones of modern and ancient continental margins, and is a standard part of the plate tectonic paradigm.

The purpose of this book is to provide a starting point for anyone interested in the background and state of knowledge concerning ophiolites (ancient oceanic lithosphere). Because ophiolites represent fragments of old oceanic crust their tectonic setting and age are extremely important in the reconstruction of ancient plate boundaries. Present-day plate tectonic theories involve the generation and disposal of oceanic lithosphere, so that these ancient fragments of oceanic lithosphere can be used directly to reconstruct conditions within the ancient oceans.

Since 1970, numerous meetings and conferences directly related to ophiolites have stimulated worldwide interest in the subject. As part of the International Correlation Program, the project "Ophiolites of Continents and Comparable Oceanic Rocks," was initiated by Dr. N. Bogdanov, Geological Institute, Moscow. This project has brought together an international group that has focused on the outstanding problems and is now producing a world map of ophiolite distribution. Participation in many of these meetings has allowed me to exchange ideas and learn considerable geologic detail on numerous ophiolite occurrences. To those colleagues who have freely exchanged data with me I should like to say: you have enriched my knowledge and provided me with insights not easily gained by pursuing the literature. It would be difficult to list each individually; however I would like to thank each one for my continuing education on ophiolites and hope that you can find parts of this book that reflect your influence on me.

At the start of this book I had hoped to compile a nearly complete bibliography of the world literature on ophiolites but in so doing realized that such a compilation would be much too large for such a short book. There are of course many references that are not listed, particularly Russian and European, mainly because of my lack of knowledge of these languages. I used the arbitrary cut-off date of December 1975 for inclusion of citations in the bibliography.

I would like to dedicate this book to my wife, Cathryn, whose continued support of my career has been a source of inspiration for me. There are many people who have helped in the preparation of this book. Those who critically read all or parts of the manuscript have my particular gratitude as they certainly helped improve the final product. My thanks to: W. Gary Ernst, Ed D. Ghent, W. Porter Irwin, Zell E. Peterman and Wayne E. Hall. Mary Donato was very helpful in the final stages of editing and proofreading of the type script so carefully produced by Laura Harbaugh. I would also like to acknowledge the drafting done by Fidelia Portillo and my son Dirk J. Coleman. This work was carried out under the supervision of Wayne E. Hall whose encouragement and friendship contributed to its successful completion. Publication of this book has been authorized by the Director of the U. S. Geological Survey.

Menlo Park, California   ROBERT G. COLEMAN
April, 1977

Contents

World Map of the Principal Ophiolite Belts Inside Back Cover

Part I. What is an Ophiolite?

Utilization of scientific terms in geology has several inherent problems, particularly when contemplating both process and description. Ophiolite has had a particularly notorious history in this respect because it has been used in various ways by numerous authors (Green, 1971). Originally, Brongniart (1827) utilized ophiolite to describe serpentinites. The name is derived from the Greek root "ophi," which means snake or serpent, because the greenish, mottled and shiny appearance of sheared serpentinite is similar to some serpents. Thus, ophiolite was employed as an alternate for serpentinite, both being more or less synonymous. Dana (1946) lists "verd antique," "ophite," "ophiolite," and "ophicalcite" as varietal names for serpentinite mixed with carbonate minerals. To further confuse the situation, Fouqué and Michel-Levy (1879) applied the term "ophitic" as a descriptive textural term for diabase (dolerite). Originally then, ophiolite was a poorly defined term that was used to describe various kinds of serpentinized ultramafic rocks. Serpentinization being a process that gives rise to a great variety of products thus led to perhaps the indiscriminate use of ophiolite as an all embracing term to include rocks associated with serpentinite, particularly where they had become mixed by tectonic processes.

European mineralogists and petrologists, each in his own way, have arbitrarily extended, limited or changed the application of ophiolite in the late 19th and early 20th centuries. This apparent confusion as to the connotation of ophiolite in the geological literature in no way diminished its popularity, and its arbitrary use continued. In a benchmark paper concerning ophiolitic zones in the Mediterranean mountain chains, Steinmann (1927) introduced a remarkable new concept of including peridotites (serpentinite), gabbro, diabase, spilite, and related rocks into a kindred relationship. In doing so, he elevated ophiolite from a rock term to a rock association term. Steinmann (1906, 1927) was impressed by the common occurrence of these kindred rocks in rootless thrust sheets of the Appennines and applied the genetic term ophiolite with the following description: "As ophiolites one must characterize only the consanguineous association of predominately ultrabasic rocks of which the principal one is always peridotite (serpentine) with subordinate gabbro, diabase, spilite, or norite and related rocks. The name ophiolite is not permissible for those rock groups that consist only of diabase-like rocks of quite similar composition and structure." Unfortunately, this definition did not specify the chemical or physical processes that were responsible for the origin of these rocks; presumably Steinmann considered the ophiolites to be formed by a consanguineous igneous process.

Steinmann (1927) describes the magmatic evolution of the ophiolite suite as follows: "First the main body of the ultrabasic mass solidifies as peridotite, then follows the gabbro (the 'Eufotide' of the Italians) and associated pyroxenitic dikes (now represented by nephrite and carcaro) and finally diabase-spilite with variolitic border zones (the 'gabbro rosso' and 'verde' of the Italians) after which come copper-bearing metaliferous veins that cut the gabbro, diabase-spilite, and the sedimentary wall rocks. The densest and volatile-poor components are thus the first to solidify, but this is not merely to be explained by gravity separation as has been assumed by Staub (1922) for the alpine serpentines. If it were, the still fluid rest-magma from which feldspathic rocks crystallized must always overlie the serpentines, whereas they actually break through from below. Much more likely, we are dealing with a differentiation independent of gravity stratification, with the lighter masses always remaining in deeper magma pockets. The second differentiation process followed in the same way, with the diabase-spilite and the mineralizing fluids remaining at depth while the gabbro rose. The relative abundance of fugitive consti- tuents in the latest member of the series is emphasized by the filling of the spilite pillow interstices and varioles with secondary minerals, quartz in the copper veins and the steatitization of the wall rocks as well as certain phenomena of contact metamorphism associated with the spilite."

Steinmann further emphasized the association of deep water sedi- ments into which the ophiolites were emplaced or injected. These abyssal sediments were predominately cherts (radiolarite), pelagic clays and Calpionella-bearing limestone. Thus the "Steinmann Trinity" consisting of serpentine, diabase-spilite, and chert was gradually combined into ophiolite and so ophiolite was to grow into a genetic rock assemblage. This led to the general concept that the ophiolites represented thick submarine magmatic extrusions emplaced in the early stages of eugeosyncline development. European geologists generally accepted the Steinmann concept for ophiolites as it appeared to adequately explain the association of mafic and ultramafic rocks within alpine orogenic zones. During this same period of time, Benson (1926) advanced his ideas concerning peridotites and serpentinites, which he considered to be plutonic in nature and intrusive into folded geosynclinal sediments within orogenic belts, and called them "alpine type." At the same time in America, Bowen (1927) brought forth his concept of fractional crystallization that provided a crystal "mush" of olivine crystals separated from a basaltic magma and lubricated by small amounts of residual magmatic liquid to allow intrusion into the upper crust. The experimental work of Bowen clearly showed that a liquid peridotite magma would need a prohibitively high temperature to exist in the earth's crust. Bowen's ideas focused attention on the problematic generation of peridotite magmas and American students tended to separate the peridotites from associated gabbros, diabase, and pillow basalts and treated the peridotites as plutonic crystal accumulations. Needless to say, the occurrences and associations of the peridotites in America and Europe were quite different and this led to controversies that still exist today. Thus the term ophiolite as originally defined by Steinmann was not successfully imported to the Americas.

Petrology in the period from 1930 to 1960 was concentrated mainly on the study of granitic and volcanic rocks and only a few researchers focused their attention on the ophiolite suite. A notable exception was the work of Hess (1938, 1955a, b) who could not reconcile the findings of Bowen's experiments with the field observations he had made on numerous mafic and ultramafic rock occurrences. The apparent lack of high temperature contact aureoles around peridotites and their almost universal partial conversion to serpentinite led Hess (1938) to propose a low temperature primary peridotite magma: "If it be supposed that the ultramafic magma contained a considerable amount of water (five percent to fifteen percent), it might thus be possible to maintain such a magma liquid at the necessary low temperature. Such a supposition concerning the water content is in keeping with the apparent facts of serpentinization which indicate that the process takes place during the latter part of the crystallization of the magma."

This concept of a hydrous peridotite magma with serpentine forming as a late mineral was discarded after Bowen and Tuttle (1949) had shown that experimentally it was not possible for a peridotite melt containing water to exist below 1000° C and little possibility that it could exist above 1000° C. It was obvious that neither the field observations nor their interpretations could be reconciled with the experimental work on the formation of peridotite. Part of the problem concerning the origin of peridotites developed because ultramafic rocks from genetically unrelated occurrences were used to augment discussions. The European geologists emphasized the close association of peridotite, gabbro, and pillow basalts (ophiolites) whereas the American geologists tended to consider the peridotites separate from associated mafic rocks. Brunn

(1940, 1960, 1961), as a result of his studies in Greece—particularly at Vourinos—emphasized the close association mentioned above and considered the ophiolite suite to have formed as thick, submarine, magmatic outpourings. Brunn considered that the outpourings were basaltic and took place along rift structures at the boundaries of geosynclinal basins (see Aubouin, 1965). Differentiation of the basaltic magma after emplacement was considered to have been responsible for the apparent stratigraphic sequence from peridotite upwards into gabbros and finally into basalts (Brunn, 1960). Thus the concept was developed that ophiolites represented early differentiated basaltic outpourings in the formation of eugeosynclines (Aubouin, 1965) and this idea was widely accepted by European geologists at that time. However, the ratio of peridotites to mafic rocks ($\sim$3:1) could not be explained by normal differention of a basaltic magma. The tendency to force a single cogenetic origin on the ophiolite suite was unfortunate as it obscured significant facts that eventually would lead to more plausible solutions.

During the past 15 years a convergence of European and American opinion concerning the origin of ophiolites has resulted from a greatly expanded geologic exchange and increased intensive study of these rocks. This, combined with the new concepts of global tectonics, has achieved a radical departure from previously held concepts.

A benchmark publication that signaled an early phase of this convergence on the ophiolite problem (Wyllie, 1967) was a collection of papers by 33 authors which provided new and interesting facts on ultramafic and related rocks. In that publication, an attempt was made to distinguish the associations of ultramafic rocks. Thus, from the specified associations given by Wyllie (1967), it was possible to focus more clearly on the ophiolite problem. An abbreviated version of these associations is given below.

1. The layered gabbro-norite-peridotite association in major intrusions. Examples: Stillwater, Great Dyke, Bushveld, Muskox and Skaergaard.
2. Ultramafic rocks in differentiated basic sills and in minor intrusions. Example: Skye.
3. Concentrically zoned dunite-peridotite association. Examples: Duke Island, Alaska, Urals.
4. Alpine-type peridotite-serpentinite association (ophiolites). Examples: Papua, Newfoundland, Cyprus, Oman.
5. Minor associates of batholithic complexes. Example: Sierra Nevada.
6. Alkalic ultramafic rocks in ring complexes. Examples: Magnet Cove, Kola Peninsula.

7. Kimberlites. Examples: South Africa, Arizona.
8. Ultramafic lavas. Examples: Canada, West Australia, South Africa.
9. Ultramafic nodules. Examples: alkali basalts from various parts of the world, i.e. Hawaii, Arizona.

The subdivisions above clearly separated for the first time the alpine-type peridotites from other occurrences which in the past had been mixed together. For instance, features in the Stillwater complex previously were used to explain relationships thought to occur in alpine-type peridotites (Hess, 1938, p. 335).

Thayer (1967), in the Wyllie volume, was the only American geologist who felt that the consanguineous relationship between peridotites and associated mafic rocks was important and proposed the "Alpine Mafic Magma Stem." He emphasized that the gabbros, diabase, and associated leucocratic rocks could be conceivably derived from a single primary peridotite magma. Thayer's insistence that the ultramafic-mafic suites of the "alpine-type" were consanguineous provided impetus to American geologists to reconsider the European ophiolite concept.

Conclusions drawn by Wyllie (1967) regarding various hypotheses on the origin of ultramafic rocks reveal that a mantle origin was favored by most investigators, but there was a wide divergence of opinion on what mantle processes were involved. Essentially, at least two magmatic processes can be invoked: (1) differentiation of a basic liquid to form an ultramafic "mush" which either forms a cumulate sequence or invades as lubricated "mush;" and (2) formation of a primary peridotite magma within the mantle which is then intruded into the crust as mush or emplaced as a solid by tectonic movement. However, the basic problem of plutonic (mantle-derived) peridotites exhibiting only embarrassingly slight contact metamorphism or none at all had not been resolved at this time. At this stage, mechanisms for the emplacement of the peridotites also remained unresolved. The concepts of plate tectonics provided a new framework into which emplacement of "alpine-type" peridotites could be considered. Nearly simultaneously, a number of papers appeared (Coleman, 1971a; Dewey and Bird, 1971; Davies, 1971; Moores and Vine, 1971; Church, 1972) advocating that fragments of the oceanic lithosphere had been thrust over or into (obducted) continental margins at consuming plate margins. Comparison of the gross lithology of the oceanic crust revealed that in general it was similar to that of the peridotite-gabbro-diabase-pillow lava sequences (ophiolites) so abundant in orogenic belts of the world. In this way, the original concepts to Steinmann (1927) came full circle, modified of course by modern petrological and plate tectonics viewpoints. Closely following the presentation of this paradigm, an inter-

the Steinmann trinity.

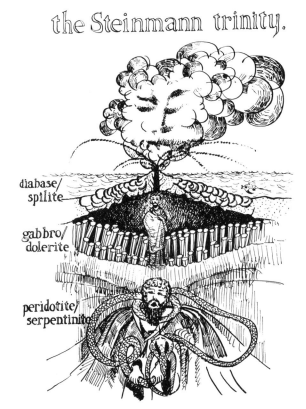

diabase/
spilite

gabbro/
dolerite

peridotite/
serpentinite

Fig. 1. Cartoon depicting a geologist's view of the Steinmann trinity. (After Prof. E. den Tex)

national field conference on ophiolites was convened under the auspices of the Geological Society of America's Penrose Conferences (Anonymous, 1972). Participants of this conference, concerned with the concept, origin, and use of the word ophiolite, produced a consensus statement that would allow its continued application but provide a tie to its earlier European utilization.

Ophiolite, as used by those present at the GSA Penrose Conference on ophiolites, refers to a distinctive assemblage of mafic to ultramafic rocks. It should not be used as a rock name or as a lithologic unit in mapping. In a completely developed ophiolite, the rock types occur in the following sequence, starting from the bottom and working up:

Ultramafic complex, consisting of variable proportions of harzburgite, lherzolite and dunite, usually with a metamorphic tectonic fabric (more or less serpentinized);

Gabbroic complex, ordinarily with cumulus textures commonly containing cumulus peridotites and pyroxenites and usually less deformed than the ultramafic complex;

Mafic sheeted dike complex;

Mafic volcanic complex, commonly pillowed.

Associated rock types include (1) an overlying sedimentary section typically including ribbon cherts, thin shale interbeds, and minor limestones; (2) podiform bodies of chromite generally associated with dunite; and (3) sodic felsic intrusive and extrusive rocks.

Faulted contacts between mappable units are common. Whole sections may be missing. An ophiolite may be incomplete, dismembered, or metamorphosed ophiolite. Although ophiolite generally is interpreted to be oceanic crust and upper mantle, the use of the term should be independent of its supposed origin.

It is difficult to ascertain whether this consensus definition of ophiolite will be widely adopted; however, recent symposia in Moscow, (Anonymous, 1973) and Paris (Mesorian et al., 1973) on this subject would suggest that the concept of ophiolites representing oceanic crust is widely accepted. Miyashiro (1973a, 1975a, b, c) has generally rejected the above definition and finds it necessary to erect a classification which seems to take us back to Steinmann's original concept. The classification of Miyashiro completely ignores the allochthonous nature of most ophiolites and their undoubted association with deep water pelagic sediments. He includes igneous rocks obviously interlayered with flysch sequences with those rocks whose origin can in no way be related to the flysch wedges. The importance of Miyashiro's classification is that it does point out that certain ophiolites have bulk chemistry different from the present day volcanics occuring at spreading centers.

It seems unlikely that there will ever be a consensus regarding the actual formation and emplacement of ophiolites. The biggest problem in characterizing ophiolites is that they may have been modified by erosion, weathering, tectonic dismemberment and/or folding as well as metamorphism. Thus, it is difficult to determine if incomplete or modified ophiolite sequences represent parts of oceanic crust or rocks produced by other geologic processes. The lack of actualistic situations that clearly record the transfer of oceanic crust onto or into the continental margins leaves the origin of ophiolites as oceanic crust up to circumstantial evidence. The main purpose of this monograph is to present the circumstantial evidence within the framework of plate tectonics.

Part II. Plate Tectonics and Ophiolites

The development of the plate tectonic hypothesis over the last decade has led to some astounding changes in geologic dogma. The long-standing ophiolite controversy with its multiple theories was one of the first petrotectonic problems to yield new solutions within the framework of the plate tectonic theory (Gass, 1963; Dietz, 1963; Hess, 1965; Thayer, 1969b; Coleman, 1971a; Dewey and Bird, 1971; Moores and Vine, 1971). The present-day paradigm considers that ophiolite forms as oceanic crust generated at midocean ridges, from whence it slowly migrates by ocean floor spreading toward continental margins—there to be subducted into the mantle. Under some circumstances at plate boundaries, slabs of oceanic lithosphere have become detached and overide (obduction) continental margins (Coleman, 1971a). The actual mechanism of ophiolite emplacement along continental margins is strongly debated, but most geologists agree that the slabs of ophiolite are allochthonous and that they originate in an environment distinctly different from where they occur today (Davies, 1971; Zimmerman, 1972; Gass and Smewing, 1973; Mesorian et al., 1973; Dewey, 1974; Coleman and Irwin, 1974). This modern view contrasts with the older debates regarding the ultramafic parts of ophiolites, which centered on the concept of their intrusion as magmas (Hess, 1938). These past concepts considered the ophiolites to represent the earliest magmatic phase of an ensialic geosyncline, which required that the ophiolites be autochthonous and interlayered with the geosynclinal sediments (Kay, 1951; Brunn, 1961; Aubouin, 1965). Even though Steinmann (1906) and Suess (1909) understood the exotic nature of the deep sea deposits (cherts) associated with the ophiolites, the possibility that they could have been tectonically transported was not pursued at that time. Reports of detailed mapping in ophiolite areas during the 1950's and 1960's began to reveal their allochthonous nature and close association with mélanges (De Roever, 1957; Gansser, 1959; Morton, 1959; Coleman, 1963, 1966; Graciansky, 1967; Decandia and Elter, 1969). Once it was demonstrated that the present-day seismic belts represent sutures where tectonic movements between rigid plates occur (Isacks et al., 1968), it was instantly appealing to call on plate interactions as mechanisms of transferring oceanic crust on or into continental margins (Coleman, 1971a; Dewey and Bird, 1970, 1971; Bailey et al., 1970). Such a hypothesis resolved the paradoxical occurrence of high-temperature ultramafic and mafic rocks within sediments that revealed no evidence of contact metamorphism. According to the hypothesis, ophiolites represent oceanic lithosphere formed chiefly at an accreting plate boundary where most of the igneous activity takes place. Following accretion at the spreading ridge, the cooled oceanic lithosphere is emplaced as a low-temperature detached slab into the sediments of the continental margin. Nearly all of the tectonic aspects of plate boundaries have actualistic present day counterparts except for the emplacement of ophiolite (oceanic lithosphere). As a result, the ophiolite-oceanic lithosphere paradigm is based largely on circumstantial evidence and must, therefore, be exposed to close scrutiny and criticism.

It is evident that if present-day oceanic lithosphere is representative of ancient oceanic lithosphere (ophiolite) then comparisons of their respective petrologic, geologic, and physical properties should reveal strong similarities. The basic flaw in this reasoning is that it is assumed that the present-day processes that give rise to new oceanic crust are the same as those that produced ophiolites in the past and can be directly compared. In making the comparisons between on-land ophiolites and present-day oceanic lithosphere several problems arise. Perhaps the most serious problem is the great age difference between present-day oceanic spreading centers and on-land ophiolites that have formed in ancient oceans now vanished. The assumption is made that the same processes producing present-day oceanic crust have been operative throughout Phanerozoic time. Geophysical measurements and estimates of the internal nature of spreading centers and resultant oceanic crust represent only a very small span of observational time, perhaps at the most 1–15 m.y. To assume that there is no secular variation of this process can only lead to complications. For instance, nearly all of the models of spreading centers have been derived from the mid-Atlantic ridge (Talwani et al., 1965, 1971; Cann, 1970, 1974; Miyashiro et al., 1970; Aumento et al., 1971), yet none of the Atlantic oceanic crust developed over the last 80 million years has been transferred onto the continental margins by plate tectonics and so we do not have available an exposed on-land section of this Atlantic oceanic crust for detailed study. Ophiolites derived from the Tethyan Ocean within the Mediterranean area are well exposed and have been under intensive study for many years (Moores, 1969; Gass and Smewing, 1973; Mesorian et al., 1973). Much of the information derived from these Tethyan ophiolites has been used to develop models for present day spreading centers (Cann, 1974). These Tethyan ophiolites are considered to represent oceanic crust formed by spreading in a Jurassic ocean which has now completely disappeared as a result of plate interactions between Africa and Europe. One could ask the question: "Were the spreading centers that produced Jurassic oceanic crust in the Tethyan Sea the same as those now forming oceanic crust within the mid-Atlantic ridge?" No doubt complications will arise when features observed from ancient on-land ophiolites are used to develop models for the hidden parts of present-day spreading centers and vice versa (see for instance, Cann, 1974, p. 184). Nevertheless, such comparisons are valid in establishing geologic models but should always be used with the utmost care.

Formation of oceanic lithosphere in other tectonic situations besides mid-ocean ridges should also be carefully considered when comparisons are being made between ophiolites and oceanic lithosphere. For instance, it is quite reasonable to expect some oceanic crust to form in marginal basins behind oceanic island arcs and presumably above subduction zones (Karig, 1971; Dewey and Bird, 1971). Unfortunately, at this time, there are not enough detailed studies on marginal basins on which to formulate a model which can adequately explain the formation of oceanic crust in this situation. Small ocean basins such as the Red Sea are presently spreading and new oceanic crust is being formed but it is not clear that the spreading processes and products are similar to those currently held for the mid-Atlantic ridge. A recent report on a Miocene ophiolite formed in the early stages of Red Sea spreading describes variations in structure and chemistry distinct from previously described models for spreading ridges (Coleman et al., 1975a). Another possible variant in the production of oceanic crust is the development of crust along leaky transform faults (van Andel et al., 1969).

Up until this time we have developed no chemical, geological, or physical means of distinguishing between these various tectonic situations within the oceanic realm. Further complications can be anticipated where there is tectonic superimposition of oceanic crust on island arc volcanics or island arc volcanics are constructed upon oceanic lithosphere (Ewart and Bryan, 1972; Miyashiro, 1975a, b, c). In these situations careful geological studies must be undertaken prior to using detailed chemical comparisons to establish the lineage of any particular ophiolite. This then returns us full cycle back to volcanics within geosynclinal sediments where the eugeosyncline is characterized by early interlayered mafic volcanics (Kay, 1951). Thick piles of submarine basalts are sometimes interlayered within the flysch sediments that develop along former continental margins (Sugisaki et al., 1971, 1972; Suzuki et al., 1972; Cady, 1975). These basalts are chemically similar to oceanic basalts but are not sequentially associated with sheeted dikes, layered gabbros, and peridotites. Obviously these volcanics cannot be related to oceanic lithosphere but represent a special type of volcanism associated with continental margins and the formation of clastic wedges. Unfortunately, Miyashiro (1975a) has classified the autochthonous submarine basalts within the continental margin clastic wedges as nonsequential ophiolites restituting the past confusion surrounding the Steinmann Trinity. I do not accept Miyashiro's ophiolite classification and consider the autochthonous basalts within these clastic wedges as an igneous rock clan separate and distinct from ophiolites.

Let us now consider the actual comparisons that have been made between oceanic crust and on-land ophiolites. Numerous seismic refraction studies at sea, recently summarized by Christensen and Salisbury (1975), provide the indirect evidence that demonstrates the rather simple layered structure of the oceanic crust. These seismic studies combined with more than 200 Deep Sea Drilling sites provide a basis for subdividing the oceanic crust into three distinct layers. The top layer variously called sedimentary layer or Layer 1, consists of sediments in various states of consolidation that are time-transgressive on the igneous crust as it moves away from the spreading center. The thickness varies with age of the crust and source of sediments within the oceanic realm. Average thickness is 0.3 km and seismic velocities (Vp) between 1.5 and 3.4 km/s (Shor and Raitt, 1969). Layer 2 is considered to be mostly submarine pillow basalt whose magnetic properties are largely responsible for the linear magnetic anomalies observed parallel to spreading ridges (Vine and Matthews, 1963; Heirtzler and Le Pichon, 1965; Cann, 1968). The average thickness of Layer 2 is 1.39 ± 0.50 km and it has an average seismic velocity (Vp) of 5.04 ± 0.69 km/s (Christensen and Salisbury, 1975). The upper parts of Layer 2 have been penetrated by Deep Sea Drilling and confirm its basaltic nature, at least within its top. Layer 3, sometimes called Oceanic layer, has an average thickness of 4.97 ± 1.25 km and an average seismic velocity (Vp) of 6.73 ± 0.19 km/s (Christensen and Salisbury, 1975). No direct sampling of the Layer 3 has been accomplished by drilling but dredge hauls have undoubtedly recovered fragments of Layer 3. At the base of Layer 3 Sutton et al. (1971) discovered a basal layer ~ 3 km thick with abnormally high velocities of 7.1–7.7 km/s within various parts of the Pacific Ocean. Layer 3 thickens by 2 km 40 m.y. after its formation within a spreading ridge and concomitantly the seismic velocity (Vp) decreases from 6.8 to 6.5 km/s approximately 80 m.y. after formation (Christensen and Salisbury, 1975). Directly under present day spreading ridges, the attenuated Layer 3 rests on anomalous mantle whose compressional wave velocities (Vp) are low (7.2 – 7.7 km/s) (Talwani et al., 1965); however, oceanic crust 15 m.y. or older rests on mantle whose Vp is 8.0–8.3 km/s (Christensen and Salisbury, 1975) see Figure 2. Below Layer 3, or Mohorovicic discontinuity, it is generally agreed that the seismically anisotropic mantle consists

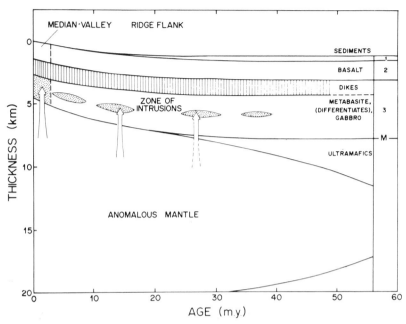

Fig. 2. Generation and evolution of the oceanic crust. The upper levels of the crust (*1, 2*) are formed largely under the median valley, but layer *3* continues to thicken for nearly 40 m.y. by intermittent offridge intrusion from the anomalous mantle. (After Christensen and Salisbury, 1975)

of a depleted peridotite that has undergone subsolidus deformation (Christensen and Salisbury, 1975). It must be understood that the seismic information has a tendency to average out irregularities within the oceanic crust as does similar information for the continental crust. This results in an oversimplified layered sequence that can be developed by reconstructed seismic profiles for any part of the earth's crust. However, continued application of more sophisticated geophysical studies of the oceanic crust will produce new data that undoubtedly will show increasing complexity within the oceanic crust.

Comparison of the oceanic crustal sequence developed mainly from oceanic seismic refraction data with the established stratigraphic sequences of on-land ophiolites shows many obvious similarities and has been used as a strong argument for correlation between them (Bailey et al., 1970; Coleman, 1971a; Dewey and Bird, 1971; Church, 1972; Mesorian et al., 1973; Peterson et al., 1974; Moores and Jackson, 1974; Christensen and Salisbury, 1975). Shortcomings of this comparison are readily apparent from Figure 3 as most ophiolites are much thinner than

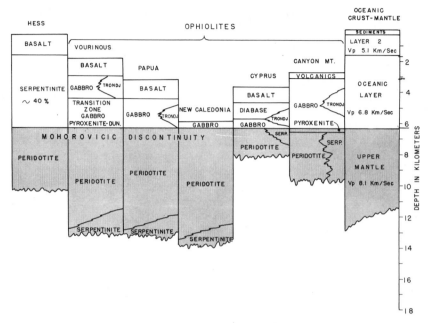

Fig. 3. Comparison of the stratigraphic thickness of igneous units from various ophiolite masses with the geophysical estimate of the oceanic crustal layers. (After Coleman, 1971a)

the average oceanic crust, however, the basalts from certain ophiolites are much thicker in some instances than the average seismic thickness of Layer 2 which is thought to be basaltic (Davies, 1971). Part of this problem can be resolved by including the cumulate ultramafic section of the ophiolites within the oceanic layer as suggested by Moores and Jackson (1974). Nonetheless, the analogy seems valid but future comparisons should be based on variations in the oceanic crust from marginal basins, small ocean basins, or deep ocean basins in relationship to the possible situation where the ophiolite may have formed. Tectonic duplication or thinning combined with erosional modification can also alter the onland ophiolites.

The stratigraphic analogy is strengthened when petrographic comparisons are made between the igneous and metamorphic rocks dredged and drilled from the oceanic crust with those rocks that make up ophiolites (Miyashiro et al., 1969; Thayer, 1969b; Moores and Vine, 1971; Coleman, 1971a; Mesorian et al., 1973; Christensen and Salisbury, 1975). Nearly all the lithologies reported from on-land ophiolites are present in oceanic basins; seismic velocities measured on these same materials can allow the rocks to be assigned to the various oceanic seismic layers as related to their petrologic character (Christensen and Salisbury, 1975). Comparisons are continuing, and, up to this time, there have been no major discrepancies perceived in the general paradigm. Later chapters will discuss the petrology of ophiolites in more detail, however, the subject of ocean crust petrology and its detailed comparisons is beyond the scope of this book.

The presence of deep ocean pelagic sediments deposited unconformably upon the upper surfaces of the ophiolite pillow lavas or interbedded with them is an important aspect and often provides the evidence of the ophiolites' allochthonous nature, particularly when the ophiolite is in tectonic contact with shallow water marine sediments (Abbate et al., 1973). The discovery that the umbers (iron-manganese sediments) deposited in small depressions within the ophiolite pillow lavas are identical to the metal-bearing sediments found on present-day oceanic crust, provides another link to the oceanic realm (Corliss, 1971; Spooner and Fyfe, 1973; Bonatti, 1975; Robertson, 1975). Related to these metal-bearing sediments is the widespread hydrothermal alteration of the oceanic crust by hot circulating sea water near spreading centers (Gass and Smewing, 1973; Spooner and Fyfe, 1973; Miyashiro et al., 1971). Alteration of the oceanic mafic rocks to zeolite and greenschist assemblages in the upper 2–3 km near spreading centers has its counterpart in the metamorphosed pillow lava and sheeted dike sequences in many ophiolites (Gass and Smewing, 1973; Bonatti, 1975, Magaritz and Taylor, 1976; Bailey and Coleman, 1975). The occurrence of massive strata-bound sulfide deposits within the ophiolite pillow lavas that are related to the hydrothermal alteration points to a strong link between oceanic processes now occurring in such places as the Red Sea (Sillitoe, 1972, 1973; Duke and Hutchison, 1974). Stable isotope fractionation within the hydrothermally altered ophiolite mafic rocks definitely shows that the ocean water was involved in their alteration and that the sulfur in the massive pyrites was a reduced ocean water sulfate (Johnson, 1972; Heaton and Sheppard, 1974; Bachinski, 1976).

There are several inconsistencies in the ophiolite-oceanic crust paradigm that require discussion and certainly provide interesting areas for additional research. As mentioned earlier in this chapter, there is no compelling evidence demonstrating that processes characterizing present-day accreting plate boundaries can be considered uniform through the Phanerozoic. Changes in spreading rate, degree of partial melting, secular changes in asthenosphere rheology and magma composition singly or together could significantly change the thickness, shape, and composition of the oceanic crust. Evidence of this is provided by the striking difference between Archean "oceanic crust" and the Phanerozoic ophiolites as demonstrated by Viljoen and Viljoen (1969), Glikson (1972) and Naldrett (1972, 1973). The chemical and petrologic differences between igneous rocks from mid-ocean ridges and ophiolites shown by Miyashiro (1973a, 1975a, b, c) require further study. This could lead to the not unlikely conclusion that oceanic crust can be developed in many tectonic situations and that basalt chemistry similar to present-day mid-ocean ridge rocks could also be generated in these other tectonic situations within the oceanic realm. The presence of sheeted dikes in most of the Tethyan ophiolites and their apparent absence, particularly in circum-Pacific ophiolite occurrences, indicate that the spreading mechanism requiring a conjugate dike system is not a universal feature of accreting margins as proposed by Moores and Vine (1971). The obvious deficiency in total thickness of ophiolite sequences compared to the average oceanic crust brings up the question posed by Smith (1974): "Up to now, it has apparently never occurred to anybody to question whether or not ophiolite sequences are actually being compared with the correct model of the ocean floor." Indeed it seems that most models of spreading ridges require knowledge derived from on-land ophiolites and so the derivative nature of the model requires more independent data or at least variations that can account for the inconsistencies. In certain occurrences of ophiolites, pelagic sediments are absent and the sedimentary sequences resting on top of the submarine pillow lavas may be continental volcano-clastic or clastic material (Bailey et al., 1970; Dewey, 1974; Hopson et al., 1975). Alternatively, the pillow lavas formed at the axis of the accreting plate margin may be covered by a thick sequence of off-axis submarine basalts of different compositions that require a parent magma distinct from that which produced the axis sequence (Gass and Smewing, 1973). Variations such as these, post dating formation of the ophiolites, indicate histories different from those permitted by current models. Some authors have commented on the short time (5–35 m.y.) spans deduced between the igneous formation of ophiolites and their tectonic emplacement (Abbate et al., 1973; Dewey, 1974; Christensen and Salisbury, 1975). The short time span between formation of oceanic crust at a spreading ridge and its tectonic emplacement indicates that most ophiolites represent oceanic crust that has been detached soon after its formation. If this relation holds for most ophiolites, tectonic emplacement is controlled by the age of the oceanic crust and old, cold oceanic crust is generally subducted and disappears in the asthenosphere. The narrow time span of ophiolite obduction has been documented by Abbate et al. (1973) who show that ophiolites in the Alps-Himalayas belt formed during the upper Jurassic were emplaced during upper Cretaceous time. Considering the volume of ophiolites tectonically emplaced during the Phanerozoic time compared to the amount of oceanic crust formed during this same period of time, less than 0.001 % of the oceanic crust has escaped subduction! It therefore seems unlikely that this small sample of oceanic lithosphere can reveal all of the secrets of the oceanic crust's evolution!

Most geological models suffer from lack of data and, as shown above the relationship between ophiolites and oceanic crust is open to question because of certain inconsistencies. The thesis of this book is not to settle these inconsistencies but to provide a source of information on the nature of on-land ophiolites.

Part III. Igneous Petrology

1. General Discussion of Ophiolite Assemblages

The most difficult aspect of the ophiolite problem is to reconcile the close association of ultramafic plutonic rocks and mafic volcanic rocks within geosynclinal sediments. In this discussion, it will be presumed that a complete ophiolite assemblage consists of the following parts: (1) metamorphic peridotite (i.e. tectonite) at the base; (2) cumulate peridotite grading upward into layered gabbros often containing plagio-granite differentiates at the top; (3) sheeted dike swarms usually consisting of 100% dikes without country rock screens and varying in composition from basalt through keratophyre; and (4) pillow lavas forming a carapace on top of the units and interlayered at the top with pelagic sediments and metalliferous precipitates. This gross stratigraphy has been clearly documented in enough places so that arguments against its existence seem pointless. On the other hand where only parts of the total ophiolite sequence are exposed within orogenic zones, a major problem arises as to its identification and origin. In such instances then, the actual geological relationships must be carefully worked out by both petrologic and structural observations. In this respect, the author will attempt to document and describe each of the separate units that make up an ophiolite with the concept that they have formed in response to processes that produced new oceanic crust.

Because all parts of the ophiolite belong to igneous ultramafic and mafic rock clans and their differentiates, the problem of their origin must be stated in terms of igneous petrology. However, the various rock units within ophiolites reveal mantle recrystallization and metamorphism, igneous cumulate sequences, differentiation products, submarine extrusion, dike injection, hydrothermal metamorphism, and tectonic deformation. All of these processes appear to be penecontemporaneous and related to the formation of the ophiolite assemblages. Therefore, in treating the sequence of events that led to the development of the ophiolite assemblage (or new oceanic crust), these events must be considered polygenetic rather than cogenetic as so often assumed for igneous rock associations or magma series. The geologic processes that reveal the polygenetic history of ophiolites can only be understood when the place and time of each separate event is established.

The oldest rocks in the ophiolite sequence are the basal metamorphic peridotites which exhibit subsolidus recrystallization at temperatures and pressures attainable only within the mantle. Direct igneous connections such as feeder dikes or transition zones between the metamorphic peridotites and the overlying cumulates are unknown. Unconformities apparently separate the two units and up to now, there has been no reliable measure of the amount of time separating these two units. The metamorphic peridotites clearly record a mantle history of subsolidus recrystallization that may have been related to a partial melting of their more labile components.

Resting upon the metamorphic peridotites are cumulate sequences that range from peridotites through gabbros to plagiogranites. Crosscutting dikes and disruption of cumulate layering indicate a dynamic situation during the development of these cumulate sequences. Source of the parent magma for the cumulate sequence is obscure. It may be the partial melt derived from the underlying metamorphic peridotite, but obvious feeders connecting these units have not been described, and the unconformity separating them attests to precumulate metamorphism and partial melting of the underlying metamorphic peridotite. Reconstruction of the cumulate sequences reveals prodigious amounts of early olivine cumulates followed by clinopyroxene, plagioclase-rich gabbros. Direct feeders of magma from the gabbro into the overlying sheeted dike-pillow lavas are not observed; however, it is not necessary to postulate numerous feeder dikes as large volumes can be extruded from just a few feeders such as we see for the Columbia River basalts.

Where the sheeted dikes and pillow lavas are exposed, their geometry suggests formation in a continuous rifting zone whereby new crust is developed within a submarine environment. The chemistry of the dikes indicates derivation from a differentiating mafic magma, but intrusive relations indicate that the underlying mafic cumulates may have solidified prior to the formation of the dike swarm. Tectonic relationships also indicate deformation of the underlying layered gabbros prior to emplacement of the dike swarms. Absence of dike swarms within certain ophiolites indicates that the tectonics of new crust formation may be varied. The pillow lavas forming above the sheeted dikes are clearly contemporaneous with the evolution of the dikes and this relationship provides the clearest evidence of distension (i.e. spreading) during their formation. Enigmatically, the evidence for distension is lacking in the underlying and presumably older cumulate sequence.

The most compelling aspect of the ophiolite sequence as described above is that fragments or screens of continental crust are not present, indicating that the formation of ophiolite thicknesses of 6–12 km must represent fragments of oceanic crust. The processes that give rise to the development of such oceanic crust involve partial melting in the mantle, differentiation of the primary liquid and extrusion of the magma under conditions of rifting or spreading.

Pervasive low grade metamorphism of the pillow lavas and sheeted dikes is characteristic of ophiolites. The metamorphic assemblages reflect mainly zeolite and greenschist facies P–T conditions and lack of penetrative deformation suggests hydrothermal metamorphism. Restriction of the metamorphism to shallow levels (1–3 km) indicates that circulation of hot ocean water near a continuous spreading center could overprint the pillow lava-dike swarms with greenschist-zeolite metamorphic assemblages shortly after their formation. Loss of varying amounts of calcium, silica, and heavy metals also provides a mechanism to develop metalliferous deposits and drastically alter the primary igneous nature of the extrusive rocks.

Chemical comparison of metasomatically altered submarine pillow lavas and sheeted dikes with fresh and unaltered oceanic and island arc volcanics is, therefore, fraught with difficulties. Present-day estimates concerning the igneous origins of ancient ophiolite volcanics have only provoked controversy and do not seem to provide definitive answers. Any partial melting of mantle material at shallow depths can provide subalkaline liquids potentially capable of producing ophiolite sequences. The difficult question to determine is: Does this process characterize only mid-ocean ridges? Our present state of knowledge indicates that unaltered volcanic rock types from island arcs, marginal basins, small ocean basins, mid-ocean ridges are present within ancient ophiolites. Based on our understanding of the polygenetic history of ancient ophiolites from the standpoint of structure, igneous processes, and metamorphism, it appears unlikely that any chemical scheme to distinguish or classify ophiolites will be practical.

Careful geological work is required to elucidate the origin of ophiolites because application of only igneous and chemical petrology to the solution of the problem could be misleading in that the effects of low-grade metamorphism must profoundly change the primary igneous mineral assemblage and the bulk rock composition. The following descriptions of the units making up ophiolites are primarily pointed towards establishing the primary igneous nature of these units. Unfortunately, there is no way of reestablishing the original bulk compositions of these rocks that the undergone low-grade metamorphism or serpentinization. Because so many of the volcanic rocks within ophiolite sequences have been called spilites, it is necessary to clarify the position of the author regarding the presence or absence of a spilitic magma as a part of ophiolite formation. The recent book by Amstutz (1974) provides the proper forum for this problem, which is beyond the context of this book. After reviewing the literature and taking into account new research on the nature of subseafloor metamorphism (Spooner and Fyfe, 1973), it seems that no spilite magma exists and that most spilites can be explained in terms of thermal metamorphism and metasomatism of normal igneous basalts.

No attempt will be made to develop an igneous classification of ophiolites as recently attempted by Miyashiro (1975a). Our present knowledge does not allow enough insight into the possible spectrum of igneous settings that could produce such an assemblage. The inherent desire of geoscientists to pigeonhole their observations into rigid classifications produces the fuel for controversies yet to come or already past. Important to this discussion is that we have similar assemblages of mafi-ultramafic rock that recur through space and time. Their geologic nature requires that they represent formation of new oceanic crust in an apparent submarine environment. Continued efforts in elucidating each occurrence both from the structural and petrologic viewpoint, will eventually allow more meaningful interpretations to be made on ophiolite occurrences.

2. Peridotites with Tectonic Fabric

2.1 Introduction

In this discussion, rocks within ophiolites consisting primarily of olivine, orthopyroxene, clinopyroxene, and spinel exhibiting tectonite fabrics are called metamorphic peridotites. Past descriptions of metamorphic peridotites were included under a broad category called alpine ultramafic rocks (Thayer, 1960, 1967). Jackson and Thayer (1972) provide a general definition: "Alpine complexes are characterized by dunite, harzburgite or lherzolite, olivine gabbro and gabbro; norite and anorthosite are rare. The rocks generally have tectonic fabrics; gneissic foliation and lineation are characteristic. Layers are lenticular, irregular, commonly discordant and tightly folded. Contact metamorphism is generally slight to obscure; many complexes are fault-bounded. Magmatic deposits consist of high-chromium to high-aluminum chromite. Alpine peridotites may be subdivided into harzburgite and lherzolite types." The main reason for publication of the Jackson and Thayer classification was to provide criteria to distinguish between the stratiform, concentric, and alpine peridotite-gabbro complexes with emphasis on the mafic-ultramafic associations. Unfortunately, in their classification the question of the ophiolite assemblage was treated as only a secondary problem and confusion remained. Earlier Den Tex (1969) porposed that peridotites be divided into nonorogenic or stratiform peridotites and orogenic or "Alpine-type" peridotites. He further divided the "Alpine-type" peridotites into ophiolitic or truly Alpine-type peridotites and orogenic "root zone" peridotites. The realization that at least some of these alpine-type peridotites were polygenetic led to a better understanding (Nicolas and Jackson, 1972). And thus the concept was conceived that the "root zone" peridotites (Den Tex, 1969) or the lherzolite subtype (Jackson and Thayer, 1972) represented a group of peridotites consisting almost entirely of ultramafic material containing higher amounts of Al_2O_3, CaO and alkalies than the harzburgite subtype associated with the ophiolites. The lherzolite subtype was further characterized by high-temperature metamorphic aureoles that seemed to provide evidence of emplacement of "high temperature" diapiric masses of undepleted continental mantle material (Green, 1967; Loomis, 1972b). Furthermore, these lherzolite subtype occurrences do not contain gabbros, diabase, or pillow lavas in a stratigraphic sequence that characterize the ophiolite sequence. The presence of lherzolite subtype masses associated with possible mantle diapirs or deep fundamental faults clearly separates the lherzolite subtype from the metamorphic peridotites associated with ophiolite sequences.

2.2 Structure

The metamorphic peridotites (harzburgite subtype of Nicolas and Jackson, 1972) occupy the basal parts of the ophiolite sequences and often are the most abundant rock type present. As described earlier from individual occurrences, the metamorphic peridotites from a substrate upon which cumulate ultramafic and gabbroic rocks are deposited (Davies, 1971; Menzies and Allen, 1974; Jackson et al., 1975). The association of undeformed cumulates with the metamorphic peridotite has been a source of confusion in previous discussions of ophiolites for several reasons.

1. There has been a tendency to separate the metamorphic peridotite (harzburgite subtype) from its associated constructional pile of ultramafic-mafic cumulates, diabase and pillow lavas.

2. Tectonic dismembering often has separated the metamorphic peridotites from the overlying constructional pile of gabbros and basalts, further reinforcing the concept that the metamorphic peridotites are not part of the polygenetic ophiolite assemblage.

Extreme tectonic dismembering of ophiolites is common within the Franciscan of California and has produced isolated "tectonic blocks" of metamorphic peridotite (Burro Mountain) within a melange (Loney et al., 1971). These isolated blocks of metamorphic peridotites originally were not considered to be parts of ophiolites. However, increased interest in the petrology of ultramafic rocks combined with the interpretation that ophiolite substrates are identical, led to the concept that these isolated blocks of metamorphic peridotites are dismembered ophiolite fragments (Bailey et al., 1970; Coleman, 1971a).

Exposures of metamorphic peridotites from ophiolites cannot be typified by any particular shape or size. The larger ophiolites such as Bay of Islands, Oman, Vourinous, Papua and New Caledonia form large tabular masses and appear to be segmented parts of a much larger continuous plate. Where ophiolites have been incorporated into orogenic zones, their peridotites become serpentinized and are often detached into isolated tectonic lenses separated from the mafic rocks. In general, gravity surveys of these lenses of metamorphic peridotites show them to have no roots and, therefore, they are not connected by "feeder pipes" to the mantle (Christensen, 1971; Thompson, 1973; Thompson and Robinson, 1975). Contrary to earlier claims concerning the nature of the contacts of the ophiolitic metamorphic peridotites, it is now generally accepted that tectonic transport was required to emplace the ophiolites and that the base of the metamorphic peridotite represents the surface of this tectonic transport (Williams and Smyth, 1973).

The internal structure of the metamorphic peridotites is generally characterized by metamorphic fabrics. Where recrystallization has been intense, foliation may be well developed and in part enhanced by segregation of olivine and orthopyroxene into separate folia. In other instances, the foliation crosses compositional layering at high angles and may be difficult to ascertain except by careful study of the orthopyroxene elongation or alinement of the spinel grains. The problem of distinguishing primary compositional boundaries versus metamorphic foliation has proved to be difficult (Loney et al., 1971; Nicolas et al., 1971). There is no doubt that primary layering has developed within the metamorphic peridotites as it is common to encounter layering manifested by varying ratios of orthopyroxene-olivine. Crosscutting bodies of dunite within the harzburgite form irregular masses which have a great range in size and diversity. Monomineralic dikes of orthopyroxenite from several centimeters up to a meter in width are characteristic. Occasional coarse-grained gabbro (ol+plg+opx+cpx) or diabase dikes may intrude the metamorphic peridotites, after deformation, but there are numerous places where such dikes are not present.

Economic deposits of chromite within the metamorphic peridotites of ophiolites have provided considerable insight into their origin and relationships with the host peridotite (Thayer, 1964, 1970; Irvine, 1967;

van der Kaaden, 1970; Engin and Hirst, 1970). A summary description has been given by Thayer (1964): "Podiform chromite deposits occur in alpine peridotite and mafic complexes and fundamentally are tabular, pencil-shaped, or irregular in form. The chromite characteristically is anhedral and commonly shows effects of granulation and magmatic corrosion. Flow layering, foliation, and lineation are parallel in most chromite deposits and peridotite host rocks, and normally pass through major rock units, including chromite; locally, foliation and lineation may cross layering. Most podiform deposits are oriented with their longer dimensions essentially parallel to layering or foliation in the host peridotite, but some are crosswise." The important aspect of the chromite deposits is that they often retain their primary structures and reveal clear evidence of having formed as cumulates from an igneous melt, in contrast to the recrystallized olivine associated with the chromite. Even though often folded, lineated, and foliated, these chromitites retain their primary cumulate structures. In contrast to the very regular and continuous chromite layers found in the Bushveld and Stillwater stratiform complexes, the podiform chromites nearly always occur as discontinuous bodies within highly tectonized and recrystallized host peridotites. The deep seated metamorphism of the peridotites tends to destroy original relations and the preserved primary structures often form high angles with structures of the host peridotite. Tectonic emplacement and serpentinization within the orogenic zones further complicates structural relationships of the podiform chromites because of the great difference in mechanical properties of host peridotites and chromitites. The chromitite bodies are usually less than 1 km in length forming tabular or lineate masses that pinch and swell into bodies up to several hundred meters and down to small irregular masses less than a meter across. Present experience shows that the podiform chromitites seem to have a random distribution in the peridotite host and most often are present within dunite masses rather than peridotite. In certain occurrences, the chromitites are spatially related to the cumulate peridotites overlying the metamorphic peridotites and may represent the base of the cumulate pile that rests upon the deformed peridotites. Future work will perhaps provide evidence that the chromites may have polygenetic origins within the ophiolites.

An important aspect of all these structures within the metamorphic peridotites is their discontinuous nature. Individual layers pinch out within several meters in most cases. In many cases isoclinal folding repeats sections and exhibits a plastic style with indistinct boundaries. The folding and compositional banding appear to have been accomplished during the high temperature and pressure mantle history of the metamorphic peridotite (Den Tex, 1969); however, these features are often obscured by nearly complete serpentinization and later shallow tectonism. Nearly all ultramafic rocks react to shallow tectonism by brittle fracture and characteristically exhibit a cubic set of joints. These joints control the invasion of water into the peridotite and develop the basic framework for the serpentinite mesh structure down to the microscopic fractures within the olivine grains. When the brittle fracture of the peridotite is accompanied by shearing and faulting, serpentinite developed along the original fractures acts as a lubricant so that movement is facilitated along these fractures. The end product of extreme tectonism and serpentinization is nearly total destruction of the original mineralogy and structures within the parent peridotite. Many of the problems in studying peridotites stem from the pervasive serpentinization and tectonism superimposed on the primary structures. A later chapter will provide more details on the problem of serpentinization.

2.3 Mineralogy and Petrography

Within the metamorphic peridotites there are two dominant rock types: harzburgite and dunite. In nearly all occurrences of ophiolites, harzburgite is the dominant rock type; however, in some instances, dunite may be predominant (see for instance, Ragan, 1963; Challis, 1965). The harzburgites generally have a xenoblastic granular texture and consist of olivine, orthopyroxene, and chromian spinel. The grain size is quite variable as many of these rocks show recrystallization textures and it is difficult to establish original grain boundaries. Orthopyroxene grains average 2–4 mm and many contain clinopyroxene exsolution lamellae. Olivine grains invariably show kink banding and as a result, original grain boundaries become obscure. The olivine grains average 2–4 mm in diameter, but it is not uncommon to find single olivine crystals up to 15 cm where the rock has not undergone complete recrystallization. The kink banding so commonly displayed in the olivine grains is also present in the orthopyroxene and clinopyroxene grains. Chromian spinel is a ubiquitous accessory mineral in the harzburgite and generally forms grains 1–2 mm in size. Green chromium diopside is also an accessory mineral in the harzburgite, but seldom exceeds 5 vol.%.

The dunites also exhibit a xenoblastic granular texture and consist predominantly of olivine with accessory amounts of orthopyroxene, clinopyroxene, and chromian spinel. Grain sizes in the dunites may be more variable than in the harzburgites and range from 2–10 mm, but also attain giant sizes up to 15 cm. Kink banding is ubiquitous in nearly all the dunites resulting in very diffuse grain boundaries.

The accessory orthopyroxene and clinopyroxene form individual discrete grains that have also undergone various degrees of kinking and bending. Chromian spinel is more abundant within the dunites, locally exceeding 4 vol.%. And as noted before, the podiform concentrations of chromitite favor the dunite as a host rock. Orthopyroxenite dikes and veins cutting the harzburgite and dunites consist primarily of orthopyroxene whose composition is similar to that in the host rocks. Occasional irregular blebs of websterite are found in the harzburgite and its minerals and textures are similar to the dunites and harzburgites.

It is worth repeating at this stage that all these rock types exhibit tectonic fabrics either in the hand specimen or microscopically. Even though banding may suggest an original cumulate origin, usually no textures remain except the podiform chromites that can be attributed to cumulus processes. There are, however, numerous recorded cases of the chromian spinels inheriting and retaining an apparent cumulate texture even though they have undergone strong metamorphic deformation (Thayer, 1964).

The modal ranges of the metamorphic peridotites from numerous ophiolites presented in Figure 4 demonstrate the restricted range in mineral ratios. There may be a continuous range in the amount of orthopyroxene and clinopyroxene from dunite to harzburgite, but the amount of clinopyroxene never seems to exceed 5 vol.% and that of orthopyroxene never exceeds 30 vol.%.

Another important point is that no systematic variation in the modal data has been noted or reported in any of the metamorphic peridotites from ophiolites considered in this study. Such widespread uniformity in these rocks indicates that their origin cannot be related directly to cumulate processes, but points more strongly to their development as a partial melting residue formed within the mantle. The lherzolite subtype or the high temperature peridotites are notably different in containing more clinopyroxene, and in some cases, plagioclase, and can be clearly distinguished from the depleted harzburgites by modal analyses, Figure 4. Where the metamorphic peridotites grade into cumulate peridotites, it is sometimes difficult to distinguish these from each other particularly where deformation was superimposed on both types during cooling of the peridotite cumulates.

Olivines from the metamorphic peridotites of ophiolites have a remarkably consistent MgO/MgO + FeO ratio (or forsterite content) (Fig. 5). Compositional zoning of the olivine has not been reported, but CO_2 fluid inclusions are present in zones within the individual crystals, but are not controlled by compositional changes within the crystal. Characteristically these olivines contain high nickel contents (0.27–0.40% Ni). It is unusual not to find at least partial sepentinization

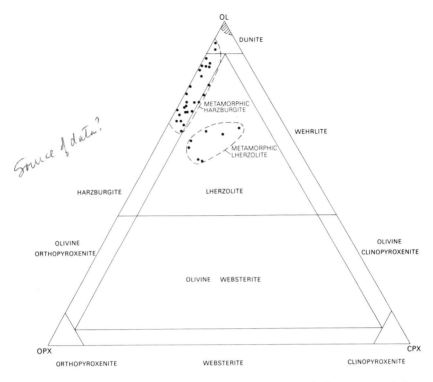

Fig. 4. Modal proportions of olivine, orthophyroxene, and clinopyroxene in harzburgite and dunite from ophiolite metamorphic peridotites compared to the lherzolite subtype

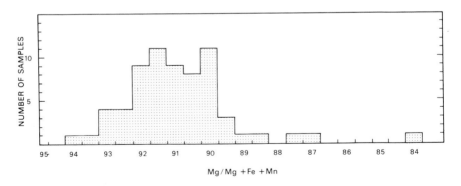

Fig. 5. Histogram showing range in composition of olivines from ophiolite metamorphic peridotites

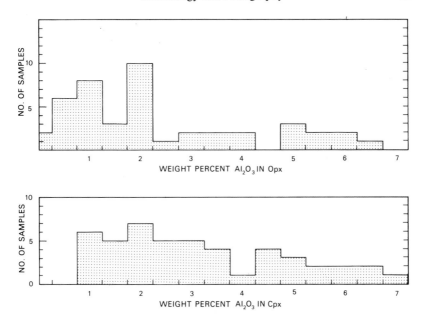

Fig. 6. Histogram showing range in Al_2O_3 contents of orthopyroxene *(OPX)* and clinopyroxene *(CPX)* from ophiolite metamorphic peridotites

of the olivine within the peridotites, and in extreme cases the olivine is completely replaced by serpentinite products. Reaction between olivine and orthopyroxene have not yet been reported from these metamorphic peridotites indicating that their equilibration was above the incongruent low pressure melting reaction relationships between olivine and ortho-pyroxene or that subsequent to the removal of partial melt, the solid assemblage achieved textural equilibrium during annealing.

The orthopyroxenes exhibit nearly identical restricted range of $Mg/Mg+Fe+Mn$ as do the olivines (Fig. 5). Variations of the alumina content are also restricted, but in some cases such as that reported for metamorphic peridotites from SW Oregon, Al_2O_3 may exceed 3% (Fig. 6) (Medaris, 1972). In most cases, the amount of chromium spinel in the rock (1–3 vol.% containing from 10–37% Al_2O_3) is insufficient to provide a high enough alumina activity to satisfy the condition of excess alumina in order to use the orthopyroxene as a geobarometer (Boyd, 1973; MacGregor, 1974). Exsolution lamellae of calcic pyroxene are often present indicating an earlier high temperature solid solution towards diopside. Plotting compositions of the orthopyroxenes on the pyroxene quadrilateral demonstrates the limited compositional ranges

Fig. 7. Clinopyroxene and orthopyroxene compositions from ophiolite metamorphic peridotites plotted on pyroxene quadrilateral

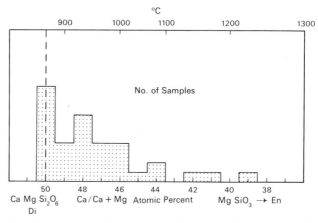

Fig. 8. Ca/Ca + Mg ratios of clinopyroxenes from ophiolite metamorphic peridotites. Temperature points taken from the diopside-enstatite solvus at 30 kbars
from Boyd (1970)

of these orthopyroxenes with a nearly constant 10 mol-% of ferrosilite
solid solution (Fig. 7). As with the olivines, no compositional zoning
has been reported in these orthopyroxenes. Serpentinization of the
orthopyroxene is usually less extensive than that of the olivines and as
a result, the orthopyroxene forms a resistant "hob nail" texture on the
weathered surfaces of the harzburgites.

The clinopyroxenes are the least abundant silicate mineral within
the metamorphic peridotites and characteristically form bright green
isolated crystals against the olive gray background of the olivine and

orthopyroxene. These clinopyroxenes are extremely calcic and contain an average of 0.67% Cr_2O_3, thereby qualifying as chromium diopsides. A plot of these chromium diopsides on the pyroxene quadrilateral again demonstrates their very restricted compositional range (Fig. 7). Plotting the Ca/Ca + Mg atomic ratio against the temperatures derived by Boyd (1970) for the diopside-enstatite solvus at 30 kb suggests that most of these diopsides equilibrated at temperatures below 1100° C and perhaps even as low as 850° C (Fig. 8). However, because these metamorphic harzburgites are so depleted in alumina, it is unlikely that pressures derived from the experimental systems can be applied to these rocks whose bulk composition is quite different than that used in the experimental systems (garnet lherzolites from kimberlites; Boyd, 1970).

The spinels in the metamorphic peridotites have extreme variations in composition when compared to their associated silicate minerals. Spinels occurring as accessory minerals in the harzburgite and dunites have Cr/Cr + Al ratios that encompass a wide range (Fig. 9). Thayer (1970) has shown that there is a bimodal distribution in the composition of chromitites from the podiform deposits. One group is characterized by high chromium and another by high alumina. A comparison with the stratiform chromitite deposits clearly shows that the main variation here is in the substitution of iron for chromium as a function of oxygen fugacities and that the aluminium-rich varieties never develop in the stratiform intrusions (Irvine, 1967). Spinels from the lherzolite subtype high temperature peridotites overlap in composition those from harzburgite subtype and provide a tie to the spinels from lherzolite inclusions found in alkali basalts (Fig. 9). Irvine (1967) has earlier pointed out that there is a positive correlation between the ratios Cr/Al and Fe^{2+}/Mg but the exchange of Cr and Al exceeds that of Fe^{++} and Mg by a factor of 2. Figure 9 further strengthens this earlier suggestion by Irvine and does seem to indicate a very restricted Mg/Fe^{2+} range within the peridotites of the harzburgite-lherzolite subtypes. The spinel composition from metamorphic peridotites is controlled largely by the Cr/Al ratio. Inspection shows that the chromium-rich spinels overlap in composition the low pressure stratiform chromitites whereas the alumina-rich spinels are more typical of the high-pressure lherzolite. The bimodal composition of the massive chromitite deposits requires some other explanation in light of the fact that the interstitial chromitite variations in Cr/Al could be explained by recrystallization reactions such as proposed by Green and Ringwood (1967, p. 156).

$$(m-2)MgSiO_3 \cdot MgAl_2SiO_6 + Mg_2SiO_4 \rightarrow MgAl_2O_4 + (m)MgSiO_3$$

aluminous enstatite   olivine   spinel   enstatite .

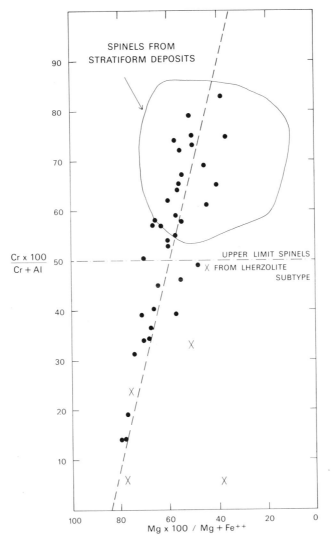

Fig. 9. A plot of accessory chromites from ophiolite metamorphic peridotites *(dots)* and lherzolite subtype *(crosses)* compared to cumulate chromites from stratiform deposits. (Modified after Irvine and Findlay, 1972)

It seems unlikely that the massive chromitites could change their compositions by subsolidus reactions as there is no evidence of such reactions present within the ore nor within experiments (Dickey and Yoder, 1972; Dickey, 1975). It appears more likely that these chemical parameters are primary igneous features. It is possible that chromite accumulations may have developed within the metamorphic peridotites at different levels within the mantle and that the high-alumina massive chromitites represent deep mantle segregations, whereas those high-chromium chromitites could represent low pressure cumulate chromite deposited during the early phases of oceanic crust formation (Dickey, 1975).

2.4 Chemistry

The metamorphic periodites have an extremely restricted range in composition again reflecting the uniqueness of their origin (Tables 1–3). In order to evaluate and compare analyses from various ophiolites, analyses were normalized after subtracting water and CO_2. A comparison of the consistency of the $MgO/MgO + FeO$ ratio for the dunite and harzburgite confirms previous suggestions that serpentinization is an isochemical process except for the introduction of water (Coleman and Keith, 1971). Thus, the normalizing processes provide a much better comparison between peridotites having varying amounts of serpentinization. The other major chemical variation related to serpentinization is the oxidization of iron, as magnetite is a by-product of this process. The ratio $MgO/MgO + FeO$ is calculated so that all Fe_2O_3 is converted to FeO. The average $MgO/MgO + FeO$ ratio for dunites is 0.86 and that for the harzburgites is only slightly lower at 0.85. The orthopyroxene and olivine have virtually the same ratio of $MgO/MgO + FeO$, so varying the proportions of these minerals does not significantly affect this ratio for the total rock. Changes in the Mg/Fe ratio reflect mainly the change in the compositions of the coexisting olivine and ortho-pyroxene. Orthopyroxene, clinopyroxene, and spinel are the only minerals containing Al_2O_3 in the harzburgite (mean 0.89% Al_2O_3) and dunite (mean 0.35% Al_2O_3). The Cr_2O_3 in the dunite and harzburgite is mainly a function of the modal amount of chromite but up to 1% Cr_2O_3 has been found in the coexisting clinopyroxenes. CaO is a minor con-stituent in both dunite and harzburgite and is contained in both clino-pyroxene and orthopyroxene. Exsolution lamellae of clinopyroxene within the orthopyroxene contain nearly all the calcium reported in orthopyroxene. Na_2O and K_2O is reported in many analyses of peridotites, but it is difficult to evaluate the accuracy of these determina-tions. Hamilton and Mountjoy (1965) report an everage of 0.004% Na_2O and 0.0034% K_2O for a variety of ultramafic rocks and Loney et al. (1971) report 0.005% Na_2O, 0.0002% K_2O for dunite and 0.012% Na_2O, 0.001% K_2O for harzburgites. These results show that the metamorphic peridotites are truly depleted in these elements and the trace amount of alkali is probably contained within the pyroxenes. NiO averages 0.31%

Table 1. Chemical compositions of typical dunites from ophiolite metamorphic peridotites

	1	2	3	4	5	6	7	8	9
SiO_2	40.6	40.5	40.8	40.0	39.7	42.1	40.0	40.4	40.8
Al_2O_3	0.49	0.24	0.38	0.36	< 0.10	1.1	0.94	0.14	0.54
Fe_2O_3	2.4	1.21	4.5	3.6	5.0	1.4	0.66	5.4	2.48
FeO	5.5	7.23	4.8	3.9	3.5	6.3	7.7	3.9	5.56
MgO	50.6	49.80	47.7	50.9	50.3	48.1	49.3	49.1	49.7
CaO	—	0.15	0.38	0.05	0.54	0.80	—	0.18	0.73
Na_2O	0.005	0.007	—	0.17	0.07	0.11	—	0.005	0.08
K_2O	0.0002	0.0012	<	0.21	< 0.29	—	—	0.002	0.03
TiO_2	0.01	0.013	0.21	< 0.03	< 0.13	0.05	0.013	0.008	—
MnO	0.11	0.11	0.02	0.11	0.11	—	0.12	0.17	0.11
S	0.02	—	0.15	—	—	—	—	0.04	—
Cr_2O_3	0.60	0.58	0.71	0.44	0.3	—	1.01	0.41	—
NiO	0.37	0.29	0.27	0.38	0.33	—	0.32	0.24	—
Total	100	100	100	100	100	100	100	100	100
MgO/MgO+FeO	0.87	0.86	0.84	0.88	0.86	0.86	0.86	0.84	0.86
Modal analyses									
Olivine	97.2	99	—	97	95	95.2	97.5	—	—
Orthopyroxene	—	trace	—	—	~ 3	0.5	—	—	—
Clinopyroxene	0.5	trace	—	—	—	0.3	trace	—	—
Spinel	2.3	trace	—	3	~ 2	3.9	2.5	—	—

H_2O (+) (−) deleted from analyses and then normalized.

(1) Burro Mt Av. 9 Dunites, Loney et al. (1971). (2) Twin Sisters DTS 1 U.S.G.S. Standard, unpublished data. (3) Red Mountain 5 Dunites, Himmelberg and Coleman (1968). (4) Vourinos 2 Dunites, Moores (1969). (5) New Caledonia 2 Dunites, Guillon and Routheir (1971); Guillon (1975). (6) Horoman 1 Dunites, Nochi and Komatsu (1967). (7) Dun Mt. 1 Dunites, Reed (1959). (8) Troodos 10 Dunites, Menzies and Allen (1974). (9) Papua 1 Dunites, Davies (1971).

Table 2. Chemical compositions of typical harzburgites from ophiolite metamorphic peridotites

	1	2	3	4	5	6	7	8	9	10	11
SiO_2	43.9	44.0	45.0	43.5	43.9	44.4	44.4	39.7	43.5	39.6	43.1
Al_2O_3	0.94	0.78	0.72	0.97	1.1	3.8	0.7	2.4	0.47	0.48	0.23
Fe_2O_3	1.34	3.00	2.1	2.6	1.3	2.2	0.4	3.9	5.4	4.9	1.00
FeO	6.55	5.50	5.8	5.5	6.8	7.1	7.6	3.9	3.2	5.3	6.50
MgO	45.9	45.3	44.4	45.7	45.2	39.6	44.7	48.1	45.7	45.2	48.4
CaO	0.6	0.5	0.87	0.63	0.59	2.5	0.7	1.0	0.77	1.0	0.55
Na_2O	0.012	0.006	0.007	0.58	0.13	0.3	0.1	0.1	0.006	0.16	0.07
K_2O	0.0010	0.004	0.08	<0.11	<0.01		0.1		0.002		0.01
TiO_2	0.01	0.016	0.01	0.03	0.07	0.15			0.01		0.02
MnO	0.12	0.13	0.14	0.11	0.01		0.13	0.11	0.15	0.10	0.10
S	0.025	—	—	—	—	—	—	—	0.019	—	—
Cr_2O_3	0.43	0.42	0.63	0.35	0.41	—	0.58	0.32	0.39	0.67	—
NiO	0.32	0.31	0.32	0.29	0.54	—	0.57	0.29	0.27	0.61	—
Total	100	100	100	100	100	100	100	100	100	100	100
$MgO/FeO + MgO$	0.86	0.85	0.85	0.85	0.85	0.81	0.85	0.87	0.85	0.82	0.87

Modal analyses

	1	2	3	4	5	6	7	8	9	10	11
Olivine	72.6	87	65	77	84	75.7	71	—	—	—	—
Orthopyroxene	23.9	13	29	19	15	16.4	27	—	—	—	—
Clinopyroxene	1.7	—	4.8		—	5.5	trace	—	—	—	—
Spinel	2.3	trace	1.6	3	1–2	2.2	8	—	—	—	—
Plagioclase	—	—	—	—	—	0.1	—	—	—	—	—

H_2O (+) (−) deleted from analyses and then normalized.

(1) Burro Mt., 8 Harzburgites, Loney et al. (1971). (2) Cazadero, PCC 1, U.S.G.S. Standard, unpublished data. (3) Red Mt., Calif., 3 Harzburgites, Himmelberg and Coleman (1968). (4) Vourinos, 3 Harzburgites, Moores (1969). (5) New Caledonia, 4 Harzburgites, Rodgers (1975). (6) Horoman, 1 Harzburgite, Nochi and Komatsu (1967). (7) Red Hills, 1 Harzburgite, Walcott (1969). (8) Bay of Islands, 22 Peridotites, Irvine and Findlay (1972). (9) Troodos, 8 Harzburgites, Menzies and Allen (1974). (10) Semail, 2 Peridotites, Glennie et al. (1974). (11) Papua, 1 Harzburgite, Davies (1971).

Table 3. Chemical compositions of lherzolite subtypes

	1	2	3	4	5	6
SiO_2	44.62	46.4	45.0	48.0	44.0	44.2
Al_2O_3	3.66	4.5	15.4	16.2	3.7	2.7
Fe_2O_3	0.59	tr	2.1	0.95	2.1	1.1
FeO	7.58	6.7	9.1	7.9	6.3	7.3
MgO	38.98	39.6	11.3	9.9	40.9	41.3
CaO	3.31	2.3	14.5	14.2	2.4	2.4
Na_2O	0.26	0.27	1.3	1.9	0.2	0.25
K_2O	0.00	0.02	0.09	< 0.01	—	0.015
TiO_2	0.14	tr	0.67	0.73	0.1	0.1
MnO	0.13	tr	0.2	0.15	0.2	0.15
Cr_2O_3	0.33	0.26	0.25	—	0.1	0.30
NiO	0.27	—	—	—	—	0.20
Total	99.87	100.05	99.91	99.93	100.00	100.02
MgO/MgO + FeO	0.83	0.86	0.51	0.53	0.83	0.83

Modal analyses

	1	2	3	4	5	6
Olivine	—	71.7	—		58	65
Opx	—	20.1	—		28	21.8
Cpx	—	5.5	60	~69	8	11.3
Spinel	—	2.0	—		1	1.5
Garnet	—		40	~ 6	—	—
Plg	—			~26	5	—

(1) Lherzolite, Ronda, Dickey (1970). (2) Peridotite, Beni Bouchera, Kornprobst (1971). (3) Garnet Lherzolite, Kornprobst (1971). (4) Garnet plagioclase pyroxenite, Dickey (1970). (5) Lherzolite, Lanzo Av. 12, Boudier (1972). (6) Estimated upper mantle, Harris et al. (1967).

in the dunites and 0.38% in the harzburgites and is mainly concentrated within the olivine and to a lesser extent in the orthopyroxene.

Comparison of the bulk composition of lherzolite subtype with the metamorphic peridotites from ophiolites demonstrates a clear distinction (Table 3). The lherzolites have a mean 3.6% Al_2O_3, 0.25% Cr_2O_3, 0.24% NiO and a lower MgO/MgO + FeO ratio 0.84. The lherzolites also appear to have a higher alkali content, but insufficient analyses are available to provide reliable analytical figures.

The data presented here provide sufficient consistent evidence to establish the metamorphic peridotites as special class of ultramafic rocks whose physical and chemical nature are unique and internally consistent and can be recognized both by their chemical and mineralogical characteristics.

3. Cumulate Complexes

3.1 Introduction

It is now generally agreed that a large part of the ophiolite sequence is made up of rock that has been derived by fractional crystallization of mafic magma to produce a sequence of ultramafic to leucocratic rocks. These sequences usually show a crude layering with peridotites at the base grading upward into increasingly feldspar-rich gabbroic rocks and finally into irregular, small, and discontinuous zones of plagiogranite. For the purposes of this discussion, I will include the peridotites and gabbroic rocks with obvious cumulate textures as well as those gabbros exhibiting a massive texture but that are obviously part of the cumulate complexes. In the past, the cumulate peridotites had been included with the metamorphic peridotites and not considered to be directly related to the layered gabbros. In recent years, the presence of undeformed cumulates has become recognized as a common feature of ophiolites (Juteau, 1970; Mesorian et al., 1973; Jackson et al., 1975). Where the ophiolites have undergone high grade dynamothermal metamorphism, recognition of the original cumulate texture of the gabbros is problematical and has led to numerous controversies (Thayer, 1963, 1971). In some instances, the amphibolites derived from cumulate gabbros have been interpreted as metamorphic aureoles related to emplacement of a peridotite magma. Enough detailed petrologic work has now been carried out to resolve some of these controversies, but continued studies will be necessary to clarify the problems of metamorphosed ophiolites. Even though cumulate sections are well displayed in ophiolites, none of those yet described compare in size or continuity with the well known non-ophiolitic stratiform complexes such as the Bushveld, Stillwater, Skaergaard, or Muskox (Jackson, 1971). Furthermore, relic cumulate textures in chromitites from the metamorphic peridotites are suggestive of perhaps deep mantle cumulate processes that preceeded the shallow cumulates of the ophiolites (Thayer, 1969a). Thus, extremely careful field work and petrography are required to elucidate the history of these rocks.

3.2 Structure

The cumulate complexes of larger ophiolite masses have a general tabular shape. Contacts with surrounding country rocks are tectonic and the upper parts typically grade into the overlying diabase (sheeted dikes), whereas the lower part of the cumulate section may rest uncon-

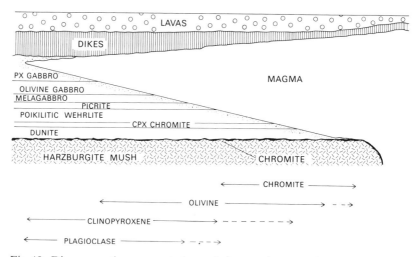

Fig. 10. Diagrammatic representation of the steady-state situation visualized beneath the axial zone of a slow spreading oceanic ridge. Crystallization relations in one half of magma chamber are shown. (After Greenbaum, 1972)

formably on the metamorphic peridotite. If one assumes that ophiolites represent oceanic crust developed at a spreading oceanic ridge, then the cumulate parts of the ophiolite represent crystal fractionation of basaltic liquids that develop into lavas and dikes at spreading centers. Greenbaum (1972) has suggested that a permanent reservoir (magma chamber) exists under a spreading ridge and that the layered sequence of an ophiolite represents cumulate phases controlled by lateral temperature gradients (Fig. 10). He visualizes a steady-state situation where the volume of magma crystallizing from reservoir matches the spreading velocity of the ridge so that the size of reservoir does not vary with time. The apparent lack of discrete magma reservoir boundaries with ophiolite cumulates would seem to support this idea. Multiple magma chambers have not yet been discovered in ophiolites but intrusive relationships within the gabbro and sheeted dike sequences indicate that more than one magma was present.

   The ultramafic parts of the ophiolite cumulates are quite variable in thickness and consist predominately of olivine with chromite horizons marking cyclic units. Various measurements of the cumulate ultramafic sections of ophiolite range from 4 km to 0.5 km in apparent thickness; however, tectonic thickening or thinning by low-angle thrusting inhibits accurate estimates of these thicknesses. Distinguishing the transition zone from cumulate peridotite to metamorphic peridotite may be complicated by post emplacement deformation and if this deformation takes place while the cumulate peridotite is only partially consolidated, it may be virtually impossible to define the contact. Development of inter-cumulate plagioclase in the peridotites and its crystal fractionation from the melt produces peridotites interlayered with plagioclase-rich gabbros, troctolites, or anorthosites. These alternating dark and light layers are easily recognized in the field and have often been referred to as the transition or critical zone (Smith, 1958; Wilson, 1959; Moores, 1969). These transition zones mark the gradual change from cumulate peridotite to cumulate gabbros and usually mark the end of bimineralic olivine and pyroxene crystallization. Various estimates of the transition zone thickness show a range from 150–750 m and must reflect the size and duration of crystal fractionation in the magma chamber. Above the transition zone, peridotites are lacking and gabbros pre-dominate. Layered gabbros sometimes develop thicknesses up to 3 km (Davies, 1971). However, in some cases layered gabbro may be absent above the transition zone; instead, only massive gabbro is present.

Detailed petrologic studies of the layered sequences in ophiolites are just beginning and it has only been recently discovered that these cumulates are cyclic and that they may represent magmatic differentia-tion sequences (Mesorian et al., 1973; Hopson et al., 1975; Jackson et al., 1975). At Vourinous, a stratiform complex nearly 1500 m thick has been recognized and its cumulate stratigraphy has been described (Jackson et al., 1975). The lower unit consists primarily of olivine cumulates with thin olivine-clinopyroxenite zones near the base that may represent beheaded cyclic units. These units are overlain by a thick olivine cumulate (500 m) containing thin chromite cumulates. A 400-m section above the predominantely olivine cumulates is characterized by cyclic unit changes that are thinner with each cyclic unit beginning with olivine cumulate and grading upward to clinopyroxene-olivine, then finally to clinopyroxene cumulates with some post cumulus plagioclase. These cyclic units are usually about 33 m in thickness and apparently the olivine cumulates decrease in thickness upward. Near the top of the Vourinous section, two-pyroxene cumulate layers form the bottom of cyclic units and grade upward into thick plagioclase two pyroxene cumulates. Jackson et al. (1975) interpret this cumulate section as a chamber slowly filling itself with magmatic sediments at or near a spreading ridge.

A noncyclic cumulate section of the Point Sal ophiolite complex, California (Hopson et al., 1975) consists of a basal olivine cumulate containing chromite (~ 150 m) which in turn grades upward into the following units. Increasing amounts of clinopyroxene form a zone of approximately 100 m of wehrlite and olivine clinopyroxenite and above this lie 150 m of cumulate clinopyroxenites, wehrlites, and olivine clinopyroxenites where olivine becomes less important as a cumulus phase. The first appearance of cumulate plagioclase in this section marks a zone 500 m thick characterized by anorthosite, troctolite, olivine-clinopyroxene gabbro, and hornblende-olivine-clinopyroxene gabbro. Above this cumulate gabbro section lie ~270 m of noncumulate clino-pyroxene and hornblende gabbros intermixed with hornblende quartz diorites, clinopyroxene-hornblende diorite, microdiorite and diabase. Hopson et al. (1975) suggest that the cumulate section represents a shallow pocket of gabbroic magma that had differentiated under a cover of submarine lavas and dikes.

The interesting aspect of these two cumulate sections from ophiolites is the difference in thickness and the non-uniform nature of the cumulate sequences. Other fragmentary information derived from the literature and by personal observation clearly points up the fact that even though the cumulate processes are an important part in the formation of ophiolites, it is readily apparent that the magma chambers of these accumulations were of transitory nature and perhaps underwent constant tectonic modification. The beheaded nature of the Vourinous units (Jackson et al., 1975) as well as field evidence of pinching and swelling of individual layered units leading to lack of stratigraphic continuity provides internal evidence of a dynamic situation during the crystal differentiation. Thayer (1963) has described flow layering in cumulate rocks from ophiolites and provides convincing evidence that at least some of these structural discordancies result from extensive flowage of crystalline mush during emplacement. Further comparisons made by Mesorian et al. (1973) of the cumulate sections from the Tethyan ophiolites demonstrates the irregularity in size and composition of these cumulate units. Nonetheless, there appears to be a general upward progression from olivine (chromite) cumulates, through alternating olivine-clinopyroxene and plagioclase-olivine-clinopyroxene cumulates and then into two pyroxene gabbros, norites, and eucrites. These layered sequences are in turn followed by massive gabbros that in many cases contain leucocratic differentiates.

3.3 Mineralogy and Petrography

The cumulate complexes are made up primarily of four minerals: clinopyroxene, orthopyroxene, olivine, and plagioclase, with important amounts of chromite in the basal sections and brown hornblende in the differentiated tops (Fig. 11). The order of crystallization of these four minerals is controlled by many factors, but perhaps the most important is composition of the parental magma and changes in its bulk

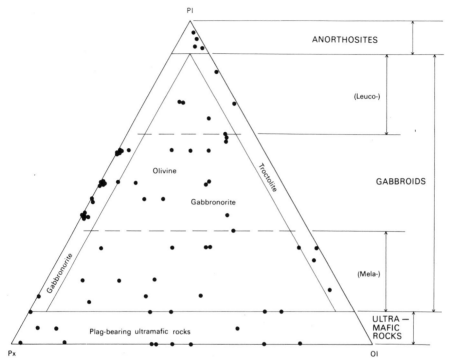

Fig. 11. Modal amounts of plagioclase, pyroxene (both ortho- and clino-) and olivine in cumulate rocks from various ophiolites. (Mostly unpublished data from U. S. Geological Survey)

composition during crystal fractionation. Up to now, no reliable estimates can be made as to the composition of the parent magmas except that they appear to have been derived from a subalkaline tholeiitic magma. Therefore, in the following discussion, no real attempt is made to relate the mineralogical and petrographic data with any particular known model of differentiation.

The olivines of the mafic cumulates are generally subhedral to euhedral grains often enlarged by adcumulus growth. They rarely show the deformation so common to the metamorphic peridotites and may be up to 3 cm diameter. The olivines within the layered gabbros are generally smaller in diameter (<10 mm) and may often be enclosed in poikilitic grains of orthopyroxene. It is extremely rare to see reaction borders of orthopyroxene on the cumulate olivine. The composition of the olivines (Fo_{90-80}) in the ultramafic parts of the cumulates overlaps that found in the metamorphic peridotites but olivines from the cumulates progressively increase in iron stratigraphically upward.

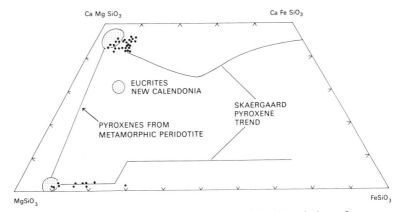

Fig. 12. Pyroxene quadrilateral showing compositional variation of pyroxenes *(black dots)* from cumulate ophiolite. Ortho- and clinopyroxenes from metamorphic peridotites are displayed along with the pyroxene trend lines from the Skaergaard intrusion. Clinopyroxenes from cumulate eucrites in New Caledonia peridotites are shown for comparison

Within the layered and massive gabbros, the olivine has a much more irregular composition and may become iron-enriched (Fo_{85-70}). Serpentinization of the cumulate olivines is widespread and often this mineral may be completely replaced by serpentine and magnetite prior to alteration of the associated plagioclase and pyroxenes.

Orthopyroxene although common in the cumulate sequences, is never abundant and does not form thick monomineralic units such as described from the large stratiform complexes (Jackson, 1971). Most commonly it forms large poikilitic (5–10 cm) grains within the ultramafic cumulates (En_{89-78}). Norites and two pyroxene gabbros may contain minor amounts of cumulate orthopyroxene (En_{80-75}) (Fig. 12). Up to now, exsolution lamellae of clinopyroxene have not been recorded from the orthopyroxenes in the ophiolite cumulate sections. Uralitization of the orthopyroxene is a common feature and actinolite and minor chlorite replace the primary crystals.

The clinopyroxenes in the ultramafic cumulates are very calcic and overlap the compositions of those clinopyroxenes from the metamorphic peridotites (Fig. 12). However, the plagioclase-rich cumulate gabbros contains slightly iron-enriched clinopyroxenes, but do not follow the compositional trend found in the Skaergaard clinopyroxenes. As the layered sequences of ophiolites become more differentiated, orthopyroxene disappears as a cumulate phase. Pigeonite has not yet been recorded in those cumulate sequences where mineral data are available. More data are required before comparisons can be made with other clinopyroxene trends, but it seems possible that the basaltic liquids differentiating to form the cumulate parts of ophiolites were perhaps richer in both magnesium and calcium. Alteration of the clino-pyroxene to green amphibole and chlorite (uralitization) is a common feature, but in some instances the clinopyroxene remains unaltered even where low-grade metamorphism has affected other coexisting igneous minerals.

Plagioclase makes an early appearance compared to the Bushveld-type, in all of the ophiolite cumulate sequences, first as an inter-cumulate phase in the cumulate dunites and then as a cumulate phase in the troctolites. The troctolites are often interlayered with clinopyroxene-olivine cumulates. Once plagioclase appears, it persists and becomes dominant in the layered gabbros, showing adcumulate textures. It is not uncommon to have monomineralic layers (anorthosite) of plagioclase interlayered with the mafic layers, particularly where there is a sharp transition from the cumulate peridotites (Bezzi and Piccardo, 1971). Characteristically the plagioclase from the troctolites and eucrites is very calcic (An_{90-95}) and rarely shows any zoning. Where stratigraphic measurements of An content have been made, systematic changes have not been established in the lower parts of ophiolite cumulate sections (Irvine and Findlay, 1972; Hopson et al., 1975). In the higher level gabbros where plagioclase is a dominant phase, zoned crystals are common and range from An_{70-80} in the core to An_{50-30} at the rim. Characteristically the plagioclase in the olivine-pyroxene rich cumulates is altered to mixtures of hydrogarnet, clinozoisite, prehnite, and chlorite as a result of serpentinization (Coleman, 1967). In the upper level gabbros, a low-grade greenschist metamorphism characteristically alters the plagioclase to albite and epidote.

Chromite may form layers within the olivine cumulates in some instances developing economic concentrations at the base of the section. Up to now, there is very little information on compositional variation of the chromite within cumulate ophiolite sections. There is some indication that the cumulate chromites have a higher Cr/Al ratio than the podiform chromites from the metamorphic peridotites.

Cumulate iron oxides are rarely reported from ophiolite sections and their lack indicates a fundamental difference from the Skaergaard which is characterized by an abundance of iron-rich layers in the upper parts of its cumulate section.

3.4 Chemistry

The cumulate rocks of ophiolite complexes show a wide range in composition and for the purposes of this discussion, are divided into two main groups: (1) ultramafic cumulates; (2) mafic cumulates (Table 4). The ultramafic cumulates are most commonly found in the

Table 4. Chemical compositions and CIPW norms of selected cumulate rocks[a]

Ultramafic cumulates

	1	2	3	4	5	6	7	8	9	10
SiO_2	37.90	39.90	40.50	39.70	43.10	44.60	39.54	42.90	44.64	45.40
Al_2O_3	8.60	10.40	9.50	3.50	1.50	2.00	5.15	18.04	17.16	12.50
Fe_2O_3	6.00	3.10	3.10	6.80	1.40	3.70	3.48	0.21	1.07	2.50
FeO	6.70	5.30	7.10	4.80	7.40	3.60	4.39	1.74	2.95	4.70
MgO	25.90	24.40	24.20	28.50	42.80	27.50	32.71	13.34	12.70	17.80
CaO	5.90	6.80	8.90	7.40	1.80	11.50	4.06	15.37	16.25	12.00
Na_2O	0.30	0.80	0.70	0.20	0.10	0.12	0.14	0.54	0.59	0.15
K_2O	—	—	—	—	0.10	0.08	0.09	0.12	0.06	0.10
H_2O^+	8.60	7.90	6.40	8.20	0.80	5.60	8.81	5.57	3.89	3.50
H_2O^-	—	—	—	—	—	—	—	—	—	—
CO_2	—	—	—	—	—	0.04	—	—	0.07	—
TiO_2	—	0.10	0.10	0.10	—	0.04	0.03	0.03	0.08	0.25
P_2O_5	0.20	—	0.20	—	0.05	—	0.20	—	—	0.01
S	—	—	—	—	—	—	—	—	—	—
Cr_2O_3	—	—	—	—	—	—	0.36	0.24	—	—
MnO	—	—	—	—	0.11	0.09	—	0.04	0.06	0.08
NiO	—	—	—	—	—	—	0.12	—	—	—
Total	100.10	98.70	100.70	99.20	99.16	98.87	99.08	98.14	99.52	98.99

Normative minerals

	1	2	3	4	5	6	7	8	9	10
Q	—	—	—	—	—	—	—	—	—	—
Or	—	—	—	—	0.6	0.5	0.6	0.8	0.4	0.6
Ab	2.8	7.5	1.6	1.7	0.9	1.1	1.3	2.3	3.3	1.3
An	24.3	27.4	24.2	9.6	3.4	5.0	14.6	50.2	46.1	34.8
Ne	—	—	2.6	0.1	—	—	—	1.4	1.0	—
Di	5.2	7.9	17.1	24.8	4.2	44.5	5.0	25.3	30.1	22.1
Hy	1.4	0.4	—	—	12.0	4.8	13.0	—	—	19.0
Ol	66.9	56.6	53.9	63.6	78.8	43.9	64.3	19.6	18.8	21.4
Cm	—	—	—	—	—	—	0.6	0.38	—	—
Il	—	0.2	0.2	0.2	—	0.08	0.06	0.06	0.16	0.5
Ap	0.5	—	0.5	—	0.12	—	0.5	—	—	0.03
MgO/MgO +FeO	0.68	0.75	0.71	0.72	0.83	0.80	0.81	0.87	0.76	0.72

[a] Norms calculated on conversion of all Fe_2O_3 to FeO, deletion of H_2O and CO_2 and normalizing

(1) Troctolite, Semail ophiolite, Oman, Glennie et al. (1974) Table 6.7. (2) Trocto-lite, Masirah Island, Oman, Glennie et al. (1974) Table 6.7. (3) Olivine gabbro, Semail ophiolite, Oman, Glennie et al. (1974) Table 6.7. (4) Cumulate peridotite, Semail ophiolite, Oman, Glennie et al. (1974) Table 6.7. (5) Gabbro, Papua ophiolite, Davies (1971) p. 26. (6) Wehrlite cumulate, Point Sal ophiolite, Bailey and Blake (1974) p. 641. (7) Wehrlite, Cyprus, Wilson (1959) p. 85. (8) Olivine eucrite, New Caledonia, Rodgers (1972). (9) Olivine-bearing uralite gabbro, Canyon Mt., Thayer and Himmelberg (1968) p. 179. (10) Olivine gabbro, Gizil Dagh, Turkey, Dubertret (1955) p. 123. (11) Gabbro, Cyprus, Wilson (1959) p. 90.

Table 4 (continued)

Mafic cumulates

11	12	13	14	15	16	17	18	19	20
46.07	46.50	46.10	47.40	48.10	48.40	50.20	50.50	53.10	55.0
22.21	7.00	15.95	27.20	16.50	19.70	16.10	18.30	18.34	15.80
0.87	2.60	5.05	0.60	0.63	0.63	1.10	0.94	2.44	4.10
2.74	6.50	1.75	1.80	3.35	3.70	4.40	5.50	7.96	7.60
8.82	21.60	10.85	3.00	11.90	9.40	11.20	9.50	5.79	3.20
17.48	11.00	16.40	15.30	16.90	16.10	12.50	12.80	9.75	9.60
0.68	0.43	0.80	2.50	0.55	0.46	0.89	1.49	0.63	0.9
0.11	0.14	0.05	0.10	0.06	0.02	0.23	0.03	0.03	0.02
1.03	2.70	1.60	1.90	1.25	1.20	2.00	0.70	1.20	1.60
—	—	—	—	—	—	—	—	—	—
—	—	—	—	0.11	—	—	0.04	—	—
0.08	0.25	0.15	0.10	0.10	0.10	0.09	0.13	0.45	1.4
0.05	0.06	0.05	—	0.02	0.02	0.03	0.01	0.03	0.09
—	—	—	—	—	—	—	—	—	—
0.04	—	—	—	—	—	—	—	—	—
0.08	0.18	0.08	—	0.08	0.10	0.10	0.13	0.17	0.12
—	—	—	—	—	—	—	—	—	—
100.26	98.96	98.83	99.90	99.61	99.83	98.84	100.07	99.88	99.43
—	—	—	—	—	—	1.1	—	12.0	18.3
0.7	0.9	0.3	0.6	0.4	0.1	1.4	0.18	0.18	0.12
4.7	3.8	7.0	17.0	4.7	3.9	7.8	12.7	5.4	7.8
57.7	17.5	41.1	64.0	43.1	52.4	40.6	43.5	47.9	40.0
0.6	—	—	2.4	—	—	—	—	—	—
23.5	31.2	34.4	10.9	32.9	22.9	18.7	16.3	0.8	7.0
—	15.9	0.05	—	9.6	19.6	30.1	25.0	32.7	23.7
12.5	30.2	16.7	4.7	8.8	0.8	—	1.9	—	—
0.06	—	—	—	—	—	—	—	—	—
0.2	0.5	0.3	0.2	0.19	0.19	0.18	0.25	0.86	2.7
0.12	0.15	0.12	—	0.05	0.05	0.07	0.02	0.07	0.22
0.71	0.71	0.63	0.56	0.75	0.69	0.67	0.60	0.36	0.22

(12) Cumulate gabbro, Red Mountain ophiolite, Bailey and Blake (1974) p. 641.
(13) Olivine gabbro, Hatay, Turkey, van der Kaaden (1970) p. 615. (14) Eucrite,
Semail ophiolite, Oman, Glennie et al. (1974) Table 6.7. (15) Olivine gabbro, Papua,
Davies (1971) p. 26. (16) Clinopyroxene-orthopyroxene gabbro, Cyprus, Coleman
and Peterman (1975) p. 1101. (17) Uralite gabbro, Cyprus, Coleman and Peterman
(1975) p. 1101. (18) High level gabbro, Papua, Davies (1971) p. 26. (19) Norite,
Vourinos ophiolite, Greece, Moores (1969) p. 46. (20) Uralite quartz gabbro,
Cyprus, Coleman and Peterman (1975) p. 1101.

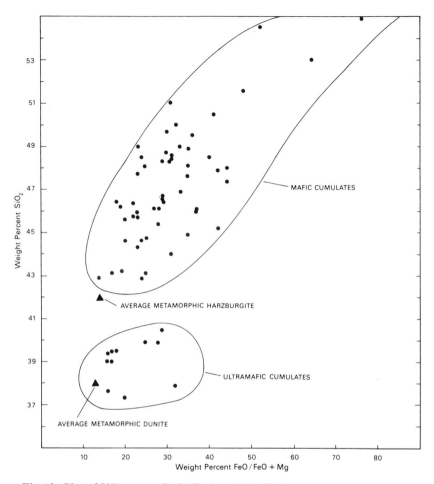

Fig. 13. Plot of SiO_2 versus $FeO\star/FeO\star + MgO$ ($FeO\star$ total iron recalculated to FeO) in ophiolite cumulate rocks. *Triangles:* average bulk composition of metamorphic harzburgites and dunites

lower parts of the cumulate sections and have $MgO/MgO + FeO$ ratios (wt %) between 0.7 and 0.8, whereas the mafic cumulates range from 0.8 to 0.2 and represent the middle and upper portions of the cumulate sections. If all available analyses of the ophiolite cumulates are plotted on a diagram (Fig. 13) with SiO_2 on the ordinate and $FeO/FeO + MgO$ (wt %) on the abscissa, it shows that the ultramafic cumulates are very similar to the dunites from the metamorphic peridotites, but that the metamorphic harzburgites appear to be different in their chemistry from nearly all of the cumulates. Additional analyses may show that this distinction may not be valid.

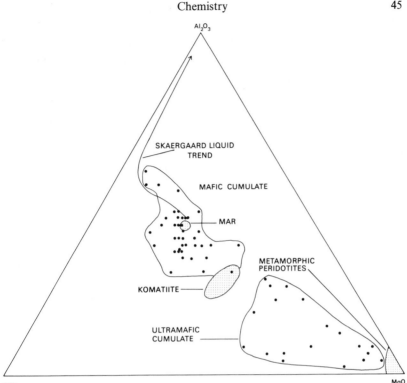

Fig. 14. Triangular diagram of MgO–CaO–Al_2O_3 in mafic and ultramafic cumulate rocks in weight percent. Komatiite field from various sources and MAR represents average composition of mid-ocean ridge basalts. Skaegaard liquid trend is displayed as a comparison that illustrates possible corrollaries to differentiation of a basaltic liquid in ophiolite sequences

The bulk of the cumulate rocks does not show a strong iron enrichment as the magma becomes progressively differentiated, indicating that perhaps the activity of oxygen is low and that most of the available iron is contained within the silicate phases. As was mentioned earlier, it is unusual to find magnetite as a cumulus or post cumulus phase in the ophiolite cumulates. A similar pattern can be seen also for titanium where it is unusual for it to exceed 1% TIO_2, whereas the Skaergaard layered rocks may contain as much as 5% TiO_2.

The wide variation in composition of these rocks is controlled mainly by the absence or presence of calcic plagioclase. Normative anorthite ranges from about 3.5 to 64 mol-% and as yet, it has not been found to enter into any sort of cyclic variations within the layered sequences. A particularly interesting trend can be seen if the cumulate analyses are plotted on a triangular diagram showing weight percent of MgO – CaO – Al_2O_3 normalized (Fig. 14). The ultramafic

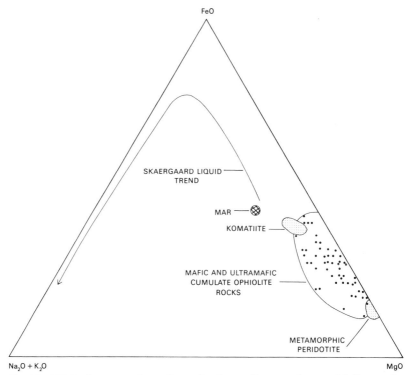

Fig. 15. AFM diagram of mafic and ultramafic cumulate ophiolite rocks. Komatiite field from various sources and MAR represents average composition of mid-ocean ridge basalts. Skaegaard liquid trend shown for comparison

cumulates occupy a restricted area characterized by only moderate amounts of Al_2O_3 and CaO, and closely approach but do not overlap the field of metamorphic peridotites. In contrast, the mafic cumulates overlap and partially follow the Skaergaard trend, but the general trend of differentiation suggests that perhaps equal amounts of calcium and aluminum deplete the magma in the earlier stages. Using this same plot, it is interesting to note where the average oceanic tholeiite and the more primitive komatiites fall on this diagram. It seems possible that perhaps the original liquid from which ophiolite layered sequences crystallized may have been similar to the komatiites and that at least some oceanic tholeiites may owe their composition to partial fractionation of a more magnesium-rich parent liquid. Further illustration of this point can be made on an AFM diagram where it is seen that cumulate rocks of the Skaergaard are quite distinct from those in ophiolite layered sequence, again suggesting a more mafic parent liquid for the ophiolites (Fig. 15). Until proportions of the various differentiates within the ophiolites are established the question of the parent liquid remains open.

Information on the minor elements within the layered sequences of ophiolites is very scarce, but the amounts of chromium and nickel are an order of magnitude less than those found in the metamorphic peridotites. However, in some transition zones of the ophiolite cumulate sections, cumulate sulfides rich in copper and nickel have been found (U.S.G.S. unpubl.).

The ultramafic cumulates differ from their metamorphic counterparts in that they contain significantly higher amounts of calcium and aluminum as well as alkalies. The presence of cumulate and intercumulate plagioclase has been used in the past to distinguish between the metamorphic peridotite and cumulate section, but it is now necessary to establish textural as well as chemical parameters carefully before the position of the cumulate section can be established.

4. Leucocratic Associates

4.1 Introduction

The close relation that leucocratic rocks have to the gabbroic parts of ophiolites and their compositional gradation from tonalite to albite granites have convinced many workers that these leucocratic rocks represent the end product of differentiation within ophiolite sequences (Wilson, 1959; Thayer, 1963; Coleman and Peterman, 1975). Where detailed mapping is available, the volume of the leucocratic rocks is small (<2%) in relation to the total exposure of the cumulate parts of the ophiolite (Wilson, 1959; Glennie et al., 1974). Clear-cut intrusive relations between these leucocratic rocks and associated mafic rocks are usually difficult to establish. However, it seems clear that the leucocratic rocks within ophiolites do not invade autochthonous rocks that are in contact with ophiolites. In many places, the leucocratic rocks interdigitate with underlying cumulate gabbros or sometimes with the overlying diabase (Wilson, 1959), and in some instances they may develop small intrusive plugs or parallel dikes within diabase dike swarms (Thayer and Himmelberg, 1968; Thayer, 1974). Keratophyre lavas may supersede basaltic extrusive rocks in some ophiolite sequences where leucocratic rocks have only rarely intruded the ultramafic parts of ophiolite sequences; however, tectonic inclusions of these rocks (many of which are altered to albitites) commonly occur within serpentinite melanges derived from dismembered ophiolites (Iwao, 1953; Coleman, 1967). There are instances where leucocratic rocks have intruded ophiolite sequences after their emplacement and in such situations careful field studies are required then to separate the leucocratic intrusives of different origins (Coleman, 1972a, p. 26–27).

4.2 Mineralogy and Petrography

Leucocratic rocks from ophiolite sequences are generally medium to fine grained and consist predominately of quartz and plagioclase with only minor amounts (usually less than 10 vol.%) of ferromagnesian minerals. Nearly all these rocks contain plagioclase feldspar that may be strongly zoned with calcic cores and sodic rims. In contrast to granophyres found in the differentiated tops of large continental mafic intrusions, the ophiolitic leucocratic rocks contain potassium feldspar only in very rare instances. Characteristically, the very small amount of potassium present in the ophiolite leucocratic rocks is contained within the plagioclase. Primary ferromagnesian minerals identified from these leucocratic rocks are most commonly hornblende or pyroxenes. Even though these rocks are low in iron, magnetite-ilmenite intergrowths are present as accessory minerals. Because these rocks represent differentiation products of a basaltic magma, their compositions range from albite granite through trondhjemite and tonalite to diorite. Available modal analyses plotted on the quartz-plagioclase-alkali feldspar triangle (Fig. 16) fall on or near the quartz-plagioclase join. Within the IUGS classification (Streckeisen, 1973), these leucocratic rocks are tonalite when biotite or hornblende is present ($> 10\%$) and trondhjemite when mafic minerals are less than 10%. When the plagioclase has an An content less than 10%, they are albite granites. Because the leucocratic rocks from ophiolites encompass a large spectrum of feldspar-quartz ratios and in the past have been given various names, Coleman and Peterman (1975) proposed the term oceanic plagiogranite as a general descriptive and collective term.

Nearly all of the oceanic plagiogranites studied by the author show the effects of low-grade metamorphism and contain epidote, chlorite, actinolite, and albite as new minerals. Metasomatism may accompany this low-grade metamorphism, but even though metamorphism and recrystallization are common in these leucocratic rocks, the modal consistency and the preservation of igneous texture indicate that the rocks are primary igneous differentiates of subalkaline basalts rather than products of metasomatism. Preservation of igneous textures and lack of metamorphic fabric suggest that the metamorphism was a static thermal event perhaps related to high-heat flow near spreading centers (Gass and Smewing, 1973).

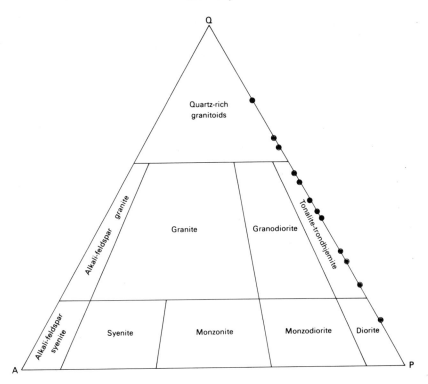

Fig. 16. Modal analyses of oceanic plagiogranites mainly from the Troodos
ophiolite, Cyprus. (Rock classification after Streckeisen, 1973)

4.3 Chemistry

High silica, low to moderate alumina, low total iron-magnesium,
and extremely low K_2O characterize the oceanic plagiogranites and their
associates (Table 5). Normative orthoclase is usually less than 4 mol-%,
and the normative An content of the plagioclase ranges from An_{21}
to An_{61}. The wide range in An content demonstrates differentiation
from basaltic composition towards leucocratic types. Plotting normative
Or, An, and Ab from these rocks on a triangular diagram reveals that
they all fall within the low pressure one feldspar field (Fig. 17). On this
same plot, O'Connor's (1965) classification places these leucocratic rocks
in the trondhjemite (keratophyre) or tonalite (dacite) field. It should be
noted that O'Connor's classification is chemical and the IUGS classification is based on modes. Comparison of the normative feldspar ratios
of oceanic plagiogranites with granophyres formed by differentiation

Table 5. Chemical compositions and CIPW norms of selected leucocratic rocks from ophiolites[a]

	1	2	3	4	5	6	7	8
SiO_2	73.6	69.4	65.4	65.2	57.5	72.5	72.6	75.8
Al_2O_3	12.3	14.0	14.5	13.5	14.5	14.0	14.4	12.9
Fe_2O_3	3.7	3.2	3.4	3.5	1.7	0.77	1.5	1.6
FeO	1.6	2.9	2.4	4.1	5.3	3.3	0.60	2.0
MgO	0.44	0.54	1.7	2.6	3.3	1.0	0.40	0.39
CaO	2.1	4.6	7.6	2.6	5.5	2.5	1.2	0.79
Na_2O	4.1	3.8	2.0	2.4	5.2	3.7	5.2	5.8
K_2O	0.33	0.07	0.30	0.64	0.20	0.33	1.1	0.20
H_2O^+	1.0	0.56	1.1	3.0	2.6	1.5	1.4 ⎱	1.0
H_2O^-	0.35	0.12	0.52	1.4	0.28	0.08	0.46 ⎰	
TiO_2	0.33	0.56	0.84	0.77	0.61	0.21	0.20	0.14
P_2O_5	0.08	0.15	0.10	0.11	0.06	0.06	0.03	0.04
MnO	0.03	0.06	0.04	0.06	0.08	0.06	—	0.06
CO_2	< 0.05	< 0.05	< 0.05	< 0.05	3.5	—	0.02	0.28
Sum.	100.	100.	100.	99.9	100.	100.	99.1	101.

Normative minerals

	1	2	3	4	5	6	7	8
Q	40.4	32.9	32.4	35.9	14.6	40.4	34.6	36.5
C	1.6	—	—	4.6	4.1	3.2	2.6	1.8
Or	2.0	0.4	1.8	4.0	1.2	2.0	6.7	1.2
Ab	35.3	32.5	17.3	21.3	45.2	31.8	45.3	49.3
An	10.1	21.2	30.0	12.8	4.9	12.2	5.9	3.7
Ne	—	—	—	—	—	—	—	—
Di	—	0.80	6.4	—	—	—	—	—
Hy	9.8	10.8	10.0	19.5	20.5	9.7	4.4	7.2
Ol	—	—	—	—	—	—	—	—
Cm	—	—	—	—	—	—	—	—
Il	0.6	1.1	1.6	1.5	1.2	0.4	0.39	0.27
Ap	0.19	0.36	0.24	0.27	0.15	0.14	0.07	0.10

[a] Norms calculated on conversion of all Fe_2O_3 to FeO, deletion of H_2O and CO_2 and normalizing.

(1) Keratophyre, Troodos Ophiolite, Cyprus, Coleman and Peterman (1975).
(2) Plagiogranite, Troodos Ophiolite, Cyprus, Coleman and Peterman (1975).
(3) Plagiogranite, Troodos Ophiolite, Cyprus, Coleman and Peterman (1975).
(4) Plagiogranite, Troodos Ophiolite, Cyprus, Coleman and Peterman (1975).
(5) Keratophyre, Point Sal Ophiolite, California, Bailey and Blake (1974).
(6) Quartz Keratophyre, Quinto Creek Ophiolite, California, Bailey and Blake (1974). (7) Quartz Keratophyre, Quinto Creek Ophiolite, California, Bailey and Blake (1974). (8) Albite granite, Canyon Mt. Ophiolite, Oregon, Thayer and Himmelberg (1968).

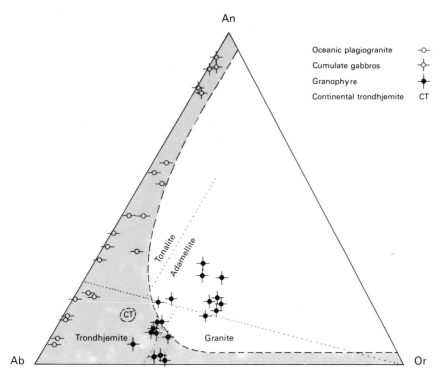

Fig. 17. Triangular diagram showing the normative content of *Ab*, *An*, and *Or* for oceanic plagiogranites, continental trondhjemite, gabbro, and granophyre. *Shaded area:* low pressure feldspar field (< 5 kbar); *dotted lines:* separate various rock types on the basis of their feldspar ratios (O'Connor, 1965)

of continental basaltic magmas shows that these granophyres all contain more than 20 mol-% orthoclase. Thus the leucocratic rocks of the ophiolites differ distinctly from continental granophyres on the basis of their normative feldspar contents.

Differentiation trends that give rise to the oceanic plagiogranites can be illustrated by the triangular AFM diagram (Fig. 18). By using the average subalkaline oceanic basalt (Troodos) as the starting liquid, it is evident that the ophiolite leucocratic rocks follow the tholeiite (Thingmuli) differentiation trend rather than a calc-alkaline trend. Perhaps the most important chemical characteristic of the ophiolite sequence is the extremely low potassium content. A semilog plot of SiO_2 versus K_2O (Fig. 19) illustrates the nature of the extremely low potassium content in these rocks compared with continental basalts

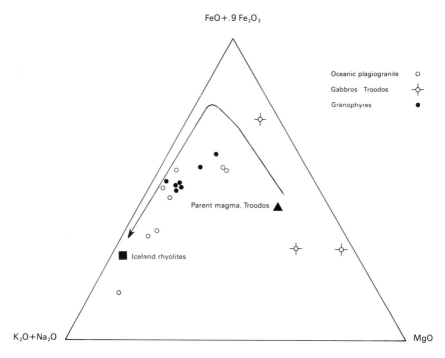

Fig. 18. AFM diagram showing the variation in oceanic plagiogranites, gabbros, and parent magma from the Troodos ophiolite, Cyprus. Granophyres and Iceland rhyolites are shown for comparison; Skaergaard liquid trend is displayed for comparison

and their differentiates. The oceanic plagiogranites contain less potassium than the continental granophyres. The $(K_2O \times 100)/(Na_2O + K_2O)$ ratio for granophyres is $\sim 49\%$, whereas that for oceanic plagiogranites is $\sim 5\%$, with no overlap between the two types.

Because low-grade metamorphism and possible metasomatism are more the rule than the exception in ophiolites, caution should be exercised in marking such comparisons. Hughes (1973) has recently made a strong case that most spilites and keratophyres are of metasomatic origin, because many of these rocks fall outside the igneous spectrum of alkali ratios. Nonetheless, there does seem to be extremely strong evidence that shallow differentiation of subalkaline tholeiites will produce leucocratic differentiates whose alkali ratios are primary but fall out of Hughes' so-called igneous spectrum. Parent melts for oceanic basalts are very poor in potassium and differentiation products will of course also be low in potassium.

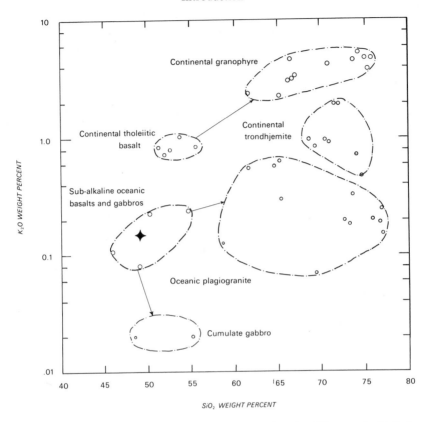

Fig. 19. Semilog plot of SiO_2 versus K_2O illustrating the difference in K_2O in oceanic plagiogranites as compared with equivalent rock types

5. Dike Swarms

5.1 Introduction

Within the upper parts of some ophiolite sequences there occur diabasic dike swarms that have developed into continuous units that may be up to 1.5 km thick. The dikes act as feeders for the overlying pillows and extend downward into the underlying gabbros. The unique feature of the dike swarms is that they consist of 100% dike rock without intervening older country rock screens. The occurrence of these remarkably continuous dike swarms has led to the idea that they have formed by continuous injection of basaltic liquid along a zone of extension (Moores and Vine, 1971). This zone of dilation could be along the axis of a mid-oceanic spreading ridge, a marginal basin, or within a small restricted ocean basin such as the Red Sea. However, not all ophiolite sequences have a distinct dike swarm in their upper constructional parts and may be superseded by massive diabase or basaltic accumulations with only infrequent dikes (Davies, 1971; Coleman and Irwin, 1974). Thus even though the dike swarms are a common feature of some ophiolites, they are by no means universal.

5.2 Structure

The internal part of the dike swarms usually consist of multiple, subparallel individual dikes that range in thickness from 10 cm to 5 m. These conjugate dikes often exhibit asymmetric chilling along their boundaries providing evidence as to their formation (Fig. 68). It is considered that the magma that forms the dikes rises along a single fracture and that it becomes chilled against previously emplaced and solidified dike. Each successive dike is emplaced within the middle of the previously emplaced dike so that repetition of this process produces a series of dikes whose chilled margins are asymmetric towards the axis of dike injection (Moores and Vine, 1971; Kidd and Cann, 1974). Statistically, the dikes from Cyprus reveal a consistent one-way chilling geometry and from this, the position and attitude of the spreading axis can be established (Kidd and Cann, 1974).

Observed field relationship reveal that the pillow lavas overlying the sheeted dikes are fed directly by these dikes. At the contact between the dike and pillow lavas, screens of pillows separate the dikes and the intrusive feeders form an irregular dendritic pattern within the pillows.

The lower contact of the sheeted dikes with the gabbros and their leucocratic differentiates is puzzling and much more complicated (Thayer, 1974). Nearly all exposures studied by the author and described in the literature provide clear evidence that the diabase dikes are chilled against the gabbro and that they also pinch-out downward into it. It would appear then that at least some of the gabbro had solidified prior to the emplacement of the dikes and that the dike intrusions went downward into the solidified gabbro. At Gizil Dagh, Turkey, a dike swarm is described that extends completely through the peridotite and gabbro indicating again a different magma source for the dikes than the gabbros and peridotites (Parrot, 1973). Deformation and possible rotation of the gabbros have been observed in several ophiolites prior to the emplacement of the sheeted dikes (Thayer, 1974). Emplacement of the sheeted dikes requires nearly 100% extension, whereas structures and dike emplacement of the underlying layered gabbros can account for no more than 10–15% extension. These features require alternate explanations for emplacement of sheeted dikes, but as yet a satisfactory solution to this puzzling structural problem has not been presented. Horizontal sheet injection is almost certainly required to explain the downward chilling into the deformed gabbros and multiple magma chambers are also necessary to provide the range in composition often observed within the dike swarm (Walker, 1975).

5.3 Mineralogy and Petrography

The rocks that make up the sheeted dike complexes of ophiolites are fine to medium-grained gray to green rocks with ophitic texture. The essential primary igneous constituents are plagioclase, clinopyroxene, magnetite or magnetite-ilmenite; occasionally, brown hornblende may supersede the clinopyroxene. Olivine and orthopyroxene have not been reported from ophiolite dike complexes. In some rocks, intersertal glass forms a groundmass containing variolitic clinopyroxene, particularly along chilled margins. Vesicles are uncommon in the dike swarms, except in some places where the dikes extend into the overlying pillows. Nearly all sheeted dike swarms that have received a careful study reveal pervasive alteration whereby the original igneous minerals have been altered to greenschist-zeolite facies assemblages. Therefore, mineralogical data on the primary igneous minerals of these rocks are extremely sparse.

The plagioclase forms as tabular crystals that provide the basic framework for the ophitic textures. Commonly, the plagioclase is strongly zoned, typically ranging from An_{55-60} to An_{20-30}. Where the dikes become more leucocratic, plagioclase dominates the rock and its rims may be An_{10-15} with the intersertal areas consisting of myrmekitic intergrowths of plagioclase and quartz. Clinopyroxene fills the gaps between the tabular plagioclase as pale brown anhedral grains. In some rocks, the clinopyroxene may show strong zoning as does the associated plagioclase. Optical and chemical data on a few clinopyroxenes from dike swarms show an iron-enrichment and a decrease in both calcium and magnesium when compared to the clinopyroxenes from the gabbros. Orthopyroxene and pigeonite have not as yet been reported from the dike swarms nor has olivine, even though olivine is present in the norm of many dikes. Hornblende reported from some dike swarms may be a late magmatic phase, but in most cases must represent low-grade thermal alteration of primary clinopyroxene. Magnetite and ilmeno-magnetite are common accessory minerals. The mineralogy of these dikes is typical of tholeiitic basalts except where advanced differentiation produces

Table 6. Selected chemical compositions and CIPW norms of sheeted dikes from ophiolites[a]

	1	2	3	4	5	6	7	8	9	10
SiO_2	50.07	53.40	49.10	58.31	49.50	51.90	58.20	45.40	48.60	51.40
Al_2O_3	15.54	15.40	15.60	13.73	14.90	15.30	15.20	15.50	18.50	15.15
Fe_2O_3	2.08	—	2.08	3.78	—	4.70	4.50	2.10	1.70	3.75
FeO	8.06	10.46	5.04	5.53	11.80	6.80	4.90	7.30	5.50	4.95
MgO	8.78	5.12	10.42	4.63	8.40	4.30	2.50	8.80	7.20	8.25
CaO	4.50	7.37	7.58	5.34	8.70	6.70	4.20	14.80	12.30	9.50
Na_2O	4.43	3.87	2.38	3.45	3.60	4.00	4.90	2.00	3.00	3.80
K_2O	0.35	0.60	1.65	0.59	0.17	0.30	0.40	0.10	0.10	0.45
H_2O	4.14	2.24	4.22	2.14	—	4.00	3.40	3.50	3.00	2.05
TiO_2	0.48	0.95	0.30	1.30	1.07	1.40	1.30	0.90	0.70	0.80
P_2O_5	0.09	—	0.04	0.11	—	—	—	—	—	—
MnO	0.20	0.12	0.18	0.25	0.20	0.10	0.10	0.10	—	—
CO_2	0.26	—	—	—	—	—	—	—	—	—
S	—	—	—	—	—	—	—	—	—	—
	99.05	99.53	98.59	99.16	98.34	99.50	99.60	100.50	100.60	100.10
MgO/MgO +FeO	0.47	0.33	0.60	0.34	0.42	0.28	0.22	0.49	0.51	0.49

Normative minerals

	1	2	3	4	5	6	7	8	9	10
Q	—	—	—	12.9	—	0.7	9.8	—	—	—
C	0.7	—	—	—	—	—	—	—	—	—
Or	2.1	3.6	10.4	3.6	1.0	1.9	2.5	0.6	0.6	2.7
Ab	39.6	33.7	21.4	30.2	31.0	35.6	43.3	8.3	23.8	32.9
An	20.7	23.5	28.7	20.9	24.4	24.1	19.1	34.1	37.7	23.5
Ne	—	—	—	—	—	—	—	4.9	1.2	—
Di	—	11.8	8.9	4.8	16.1	9.2	2.3	34.1	20.4	20.2
Hy	15.3	25.5	12.7	24.8	1.8	25.7	20.5	—	—	1.9
Ol	19.4	0.03	17.3	—	23.7	—	—	16.1	15.0	17.2
Il	1.0	1.9	0.6	2.6	2.1	2.8	2.6	1.8	1.4	1.5
Ap	0.2	—	0.1	0.3	—	—	—	—	—	—

[a] Norms calculated on coversion of all Fe_2O_3 to FeO, deletion of H_2O and CO_2 and normalizing.

(1) Greenstone, Cyprus, Bear (1960) p. 51. (2) Microdiorite, Cyprus, Bear (1960) p. 51. (3) Microgabbro, Cyprus, Gass (1960) p. 83. (4) Quartz diabase, Cyprus, Wilson (1959) p. 69. (5) Altered dolerite, Cyprus, Moores and Vine (1971) Table 3. (6) Diabase, Oman, Glennie et al. (1974) Table 6.8. (7) Spilite dike, Oman, Glennie et al. (1974) Table 6.8. (8) Hornblende diabase, Oman, Glennie et al. (1974) Table 6.8. (9) Gabbro porphyrite, Masirah Island, Glennie et al. (1974) Table 6.8. (10) Dolerite, Pinde, Greece, Rocci (1973) Table 1. (11) Dolerite, Taurus, Turkey, Rocci (1973) Table 1. (12) Dolerite, Mamonia, Cyprus, Rocci (1973) Table 1. (13) Diabase,

Table 6 (continued)

11	12	13	14	15	16	17	18	19	20
49.66	48.20	49.40	49.70	49.32	49.50	51.37	48.80	49.80	50.05
14.63	15.40	14.60	14.10	16.95	16.30	15.28	13.85	16.10	13.35
—	8.23	0.80	2.80	2.40	3.50	1.66	5.12	1.90	3.45
11.10	—	7.90	7.20	5.25	5.65	7.55	8.08	6.70	8.65
6.36	8.89	8.50	7.40	9.15	8.10	9.68	7.47	7.10	8.45
9.01	6.81	9.70	8.60	10.90	7.10	4.91	5.89	8.95	7.40
3.92	4.73	2.90	3.40	1.55	3.70	3.32	3.37	4.00	3.55
0.24	0.09	0.10	0.10	0.65	0.25	1.67	0.63	0.15	0.10
2.98	5.40	2.90	1.30	0.70	0.70	3.44	3.99	3.05	3.98
1.25	1.13	0.80	1.00	0.85	0.75	0.32	1.79	1.05	1.20
—	—	0.10	0.10	0.02	0.12	0.04	0.20	0.30	0.25
0.19	0.15	0.02	0.02	0.18	0.08	0.24	0.33	0.15	0.25
—	—	—	—	—	—				
—	—	—	—	—	—				
99.34	99.03	97.72	95.72	97.92	95.75	99.48	99.52	99.25	99.25
0.51	0.55	0.50	0.43	0.55	0.48	0.52	0.37	0.46	0.42

11	12	13	14	15	16	17	18	19	20
—	—	—	—	—	—	—	—	—	—
1.5	0.6	0.6	0.6	3.9	1.6	10.3	3.9	0.9	0.6
34.4	38.0	25.9	30.6	13.5	33.1	29.3	30.0	35.2	31.2
22.4	22.1	28.0	24.3	38.5	28.6	22.8	21.9	26.6	21.0
—	2.8	—	—	—	—	—	—	—	—
19.9	11.6	18.1	16.7	13.8	6.3	1.9	6.2	14.5	12.8
0.5	—	16.6	17.0	24.9	13.6	17.5	20.8	3.8	18.1
18.7	22.6	8.9	8.5	3.4	15.0	17.5	13.1	16.2	13.4
2.5	2.3	1.6	2.0	1.7	1.5	0.6	3.6	2.1	2.4
—	—	0.3	0.3	0.05	0.3	0.1	0.5	0.7	0.6

Bay of Islands, Newfoundland, Williams and Malpas (1972) Table 1, p. 1224. (14) Diabase breccia, Bay of Islands, Newfoundland, Williams and Malpas (1972) Table 1, p. 1224. (15) Dolerite, Gizil Dagh, Hatay, Turkey, Dubertret (1955) p. 124. (16) Dolerite, Gizil Dagh, Hatay, Dubertret (1955) p. 124. (17) Diabase, Ergani, Turkey, Bamba (1974) p. 302. (18) Diabase, Ergani, Turkey, Bamba (1974) p. 302. (19) Diabase, Fethiye Complex, Turkey, van der Kaaden (1970) p. 515. (20) Diabase, Fethiye Complex, Turkey, van der Kaaden (1970) p. 515.

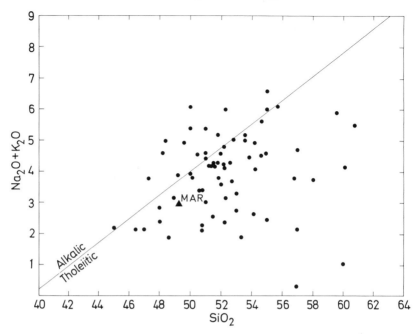

Fig. 20. $Na_2O + K_2O$ versus SiO_2 diagram of diabase in dike swarms from ophiolites. MAR average mid-Atlantic ridge basalt. All plotted in weight percent after subtracting total water and normalizing

plagiogranite (keratophyre) and the main constituents of these rocks are plagioclase and quartz as described in the previous section.

The alteration assemblages of the dike swarms are more characteristic of these rocks and typically consist of albite, chlorite, actinolite, epidote, sphene and carbonate. The calcic plagioclase usually breaks down to cloudy or dirty albite and the clinopyroxene becomes uralitized (replaced by a fibrous green actinolite). In some instances prehnite or pumpellyite will appear with these alteration assemblages. A discussion of this alteration will be given in the chapter on metamorphic petrology.

5.4 Chemistry

There are numerous chemical analyses of rocks from ophiolite dike swarms; however, the value of such analytical data is diminished since many of these rocks have been altered after their igneous emplacement. There is a general lack of descriptive petrographic data accompanying the analyses that could allow some evaluation regarding the degree of alteration or its products. The comparisons to be discussed, therefore,

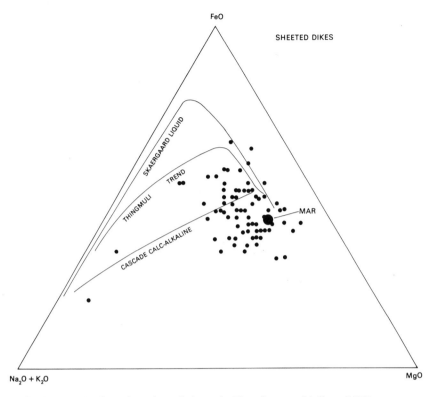

Fig. 21. AFM triangular plot of sheeted dikes from ophiolites. MAR average mid-Atlantic ridge basalt. All Fe converted to FeO* and normalized against weight percent MgO and $Na_2O + K_2O$

could be misleading as it seems probable that at least some of these rocks have undergone considerable metasomatism. This metasomatism is reflected by the very irregular nature of the normative minerals in rocks that uniformly consist of modal clinopyroxene, plagioclase, iron-titanium oxides (Table 6). A $Na_2O + K_2O$ versus SiO_2 diagram for collected analyses of sheeted dikes demonstrates that most of these rocks can be called tholeiitic basalts using the MacDonald and Katsura (1964) dividing line (Fig. 20). Those few analyses plotting above the line probably represent post igneous metasomatism rather than variation in the primary magma producing these dikes. An AFM plot of the same collected analyses provides another comparison showing that the main trend of these dike swarms is toward a tholeiitic trend rather than calc-alkaline (Fig. 21). Those points lying outside the tholeiite field are usually enriched in both sodium and potassium, suggesting

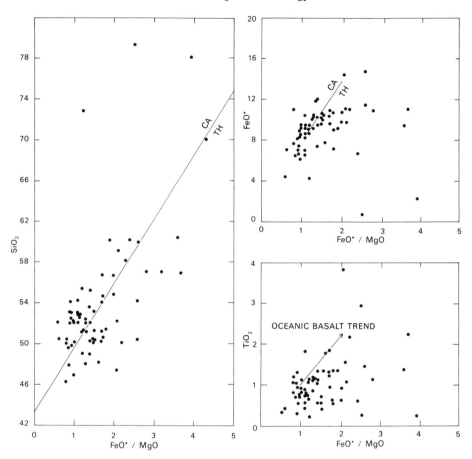

Fig. 22. Compositions of dike swarms in terms of the variation of SiO_2, $FeO^\star$ (all iron as FeO) and TiO_2 with increasing $FeO^\star/MgO$ in weight percent. *CA* calcalkalic; *TH* tholeiitic. (After Miyashiro, 1975a)

post-igneous alteration. Miyashiro (1975a) has devised a series of plots whereby the variation of $FeO^\star/MgO$ ($FeO^\star$ total iron as FeO) plotted against SiO_2, $FeO^\star$, and TiO_2 provide a way of distinguishing tholeiitic from calc-alkali basaltic rocks (Fig. 22). Application of this scheme to the collection of dike swarm analyses produces contradictory evidence when compared to Figures 20 and 21, in that many of these rocks, according to Miyashiro (1975a), would be calc-alkaline rather than tholeiite. Comparison of analyses from individual ophiolite masses does not produce any convincing trends that mimic either the calc-alkaline or tholeiitic trends provided by Miyashiro (1975a). Such comparisons are only valid when the analyzed rocks represent igneous events and not a later alteration or metamorphism. There is no criterion by which one can choose altered rocks to insure that their bulk chemistry represents original igneous compositions. Continued use of variation diagrams to provide definitive data on the igneous origin of altered and metamorphosed rocks from ophiolite sequences can only lead to unresolvable controversies, as demonstrated by Figures 21 and 22. Another feature of the chemistry of the sheeted dikes is the unmistakable enrichment of silica associated with increase of both modal plagioclase and quartz. Leucocratic dikes within dike swarms record intrusion of plagiogranite differentiates, and inspection of the AFM diagram shows a continuum between basaltic compositions and that of keratophyre (plagiogranite) (Fig. 21). The sheeted complexes of ophiolite sequences contain mafic and leucocratic dikes that have cross-cutting and chilled contact relationships that indicate separate magma sources. Chilled contacts of basaltic dikes against keratophyres (plagiogranite) and the reverse suggest separate and distinct magma chambers existed rather than a single chamber that followed a normal differentiation trend.

The weight percent of alkalis show a wide variation within the sheeted dikes (Fig. 23) and this variation must reflect a certain amount of postigneous metasomatism. Na_2O percentages have a bimodal distribution extending across the total range for abyssal tholeiites, island arc tholeiites, and oceanic island tholeiites. It is probable that the sheeted dikes having 2–2.5% Na_2O may represent primary igneous values, and those ranging from 3% to 5% have undergone post-igneous metasomatism. The K_2O contents of the sheeted dikes exhibit a maximum of 0.1–0.3 K_2O, very nearly the same as abyssal tholeiites, but the values as high as 2% K_2O must surely represent some form of potassium metasomatism especially inasmuch as no flow rocks exhibit such high potash contents.

The TiO_2 content of the collected analyses of sheeted dikes shows much less spread in values than other elements, and, therefore, may be representative of the primary igneous values (Pearce and Cann, 1971) (Fig. 24). The average content of TiO_2 for the collected sheet dikes is 0.5–1.5%, very similar to that found for abyssal tholeiites and island arc tholeiites, but much lower than the range of TiO_2 found in continental tholeiites (Fig. 24).

The sheeted dikes from ophiolite complexes appear to be characteristically tholeiitic in composition, but post-igneous alteration, combined with differentiation of the parent liquid prior to injection, has obscured the nature of the primary magma and has led to numerous controversies. Future work should concentrate on the nature of the

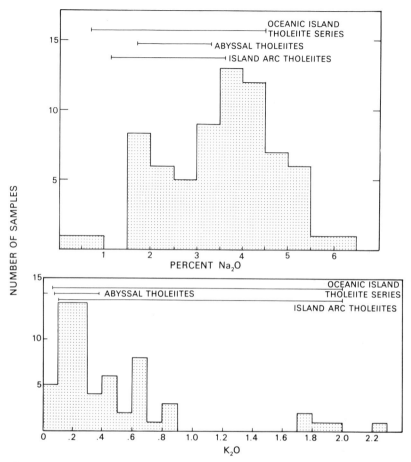

Fig. 23. Frequency distribution of Na_2O and K_2O (weight %) in dike swarms compared to oceanic island tholeiites, abyssal tholeiites, island arc tholeiites (Miyashiro, 1975a)

alteration and its effect on the igneous composition. Careful search for unaltered rocks in sheeted dikes for analytical work could lead to definitive answers on the igneous evolution of these rocks. The existence of unaltered clinopyroxenes within the sheeted dikes could also allow comparisons of their compositions by electron microprobe techniques and thus overcome the inherent problems of comparing analyses of the total rock that have undergone low-grade hydrothermal metamorphism. Present attempts to classify and distinguish ophiolites by their supposed uniqueness has failed because the chemistry presented represents both igneous rocks and their low-grade metamorphosed equivalents.

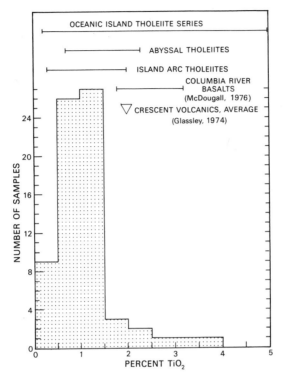

Fig. 24. Frequency distribution of TiO_2 (weight %) in dike swarms compared to oceanic island tholeiites, abyssal tholeiites, island arc tholeiites (Miyashiro, 1975a)

6. Extrusives

6.1 Introduction

Submarine pillow lavas most often form the uppermost unit of ophiolite sequences and provide direct evidence that they developed under water. It is now well known that the ocean floors are covered with pillow lava and that formation of the lava is the result of chilling liquid lava in water (Moore, 1975). Direct observation in present-day active mid-ocean ridges show that pillow lavas form the upper parts of the newly formed oceanic crust. Using this analogy, the pillow lavas within ophiolite sequences may also have formed at a spreading ridge within some ancient oceanic realm. Sediments inter-layered or resting directly on top of the pillows are almost invariably pelagic and provide additional evidence that the pillow lavas represent submarine extrusions (Robertson and Hudson, 1973; Berger, 1974;

Robertson, 1975). In some instances keratophyric lavas without pillow structures form the upper parts of ophiolites and in these situations, there is no direct evidence that these more silicic lavas formed as submarine lavas. In numerous exposures of ancient pillow lavas along continental margins or within orogenic belts, only the pillow lavas are exposed and their basement remains hidden. In these cases, much controversy has been generated regarding their origin as oceanic crust (ophiolites), as eugeosynclinal volcanics, or as island arc volcanics (Snavely et al., 1968; Sugisaki et al., 1972; Glassley, 1974; Cady, 1975; Lyttle and Clarke, 1975). Up to now there is no reliable method either chemical or structural to establish the volcanic setting of pillow lavas where the outcrops do not provide sufficient exposure of associated sedimentary or igneous rocks.

6.2 Structure

Where the pillow lavas of ophiolite masses have been mapped in detail, they appear to be tabular units (Wilson, 1959; Bear, 1960; Glennie et al., 1974). It is unusual to distinguish individual flow units except where pelagic sediments mark a hiatus in extrusion (Robertson, 1975). The total thickness of ophiolite pillow lavas show wide variation from 0.3 to 6 km, but excluding Papua which is 4–6 km thick, an average of about 1 km is obtained (Moores and Jackson, 1974). On Troodos, two chemically distinct pillow units have been defined and these units are now considered to represent separate volcanic settings (Gass and Smewing, 1973; Pearce, 1975). Within pillow sequences, there are numerous intrusive rocks that form dikes, sills, and sometimes irregular intrusive masses. Agglomerates and breccias are uncommon as are tuffaceous units. Hyaloclastites may develop between individual pillows and irregular carbonate and chert masses often occupy the voids between pillows. Normally the pillows of ophiolite sequences are tightly packed and contain very little interpillow matrix material. Moore (1975, p. 276) has observed pillows forming in underwater eruptions and describes this process as follows: "Most of the actively growing pillowed lava flows examined by divers, as well as pillowed flows in the deep sea studied from photographs or by observations from submarines, are composed of elongate, interconnected flow lobes that are elliptical or circular in cross section. The flow lobes are fed from upslope by larger, connected lava tubes that maintain lava pressure within the growing lobes. Isolated pillow sacks are rare but may form when a flow tube pinches off on a steep and tumbles a meter or so downhill."

Many of the features observed by Moore can be seen in the pillow lavas of ophiolites and bear out his proposed mechanism of formation. Where the pillow lava sequence is complete, the basal portion merges into the sheeted dike swarm in a gradual fashion whereby the dikes increase downward until there remains no interdike pillow lavas. Upward, interbedded pelagic sediments and chemical precipitates (umbers) mark the transition from igneous extrusion to a sedimentary regime. Recognition of these distinct top and bottom features in dismembered ophiolites can allow interpretation of the structural attitude of isolated blocks.

6.3 Mineralogy and Petrography

Considerable information has been developed on the nature of ophiolite pillow lavas, but there is still controversy as to the meaning of these results. In particular, many pillow lavas are spilitic in mineralogy and their bulk composition suggests that either there has been metasomatism or that the spilite composition represents a special sodium and water-rich basaltic magma (for complete discussion see Amstutz, 1974). There seems little doubt that most ancient pillows have undergone at least some chemical modification since their initial chilling and consolidation on the ocean floor. The later section on hydrothermal ocean floor metamorphism will describe some aspects of this alteration, but for the purposes of this discussion, primary igneous features will be highlighted.

Nearly all pillows show a chilled rim which may be 0.5–4 cm thick. Within the chilled margin glassy material predominates and may contain variolites, microlites, or other manifestations of incipient crystallization. Plagioclase and clinopyroxene are the most common crystallites in the chilled margin. This same glassy material commonly form a hyloclastic matrix around the individual pillows. It is rare to see glassy material preserved in the older ophiolite pillows as it is very susceptible to alteration to smectite and chlorite (Bischoff and Dickson, 1975). Igneous textures are preserved in nearly all cases within pillow lavas. Preservation of chilled margins are clear evidence of the nondeformational nature of the alteration of these rocks. Subophitic and intersertal textures develop within the pillows away from the chilled margins. Plagioclase An_{40-65}, sometimes zoned, and subcalcic clinopyroxene are the predominant minerals within the central parts of pillows. Ilmenomagnetite is a ubiquitous accessory mineral in nearly all occurrences. Scattered phenocrysts of olivine, but rarely hypersthene, are found in the pillow lavas. Where the magma has differentiated towards keratophyre, plagioclase may make up more than 80%

Table 7. Selected chemical compositions and CIPW norms of extrusive rocks from ophiolites[a]

	1	2	3	4	5	6	7	8	9	10
SiO_2	51.71	47.80	52.80	50.90	45.04	54.70	49.50	54.60	55.00	59.00
Al_2O_3	14.70	15.90	14.40	15.20	7.13	14.50	13.60	14.70	14.10	14.40
Fe_2O_3	1.86	—	—	—	3.13	4.15	5.80	3.30	4.30	5.00
FeO	6.34	10.90	8.40	13.50	5.43	6.40	6.70	4.30	2.60	3.50
MgO	7.55	8.40	7.60	5.80	26.16	5.45	7.80	8.70	7.20	2.40
CaO	10.74	9.20	9.90	8.60	5.56	9.65	8.40	3.40	7.30	3.70
Na_2O	1.88	1.50	1.80	2.60	0.77	2.80	3.45	1.20	2.80	5.90
K_2O	0.25	0.50	0.19	0.16	0.06	0.09	0.11	2.00	0.40	—
H_2O	3.70	—	—	—	3.81	0.94	2.75	7.00	5.60	3.90
TiO_2	0.48	0.86	0.52	1.32	0.36	0.70	1.24	0.30	0.30	1.30
P_2O_5	0.06	—	—	—	0.12	0.08	0.11	0.03	0.03	0.33
MnO	0.15	0.10	0.11	0.16	0.14	0.10	0.12	0.08	0.07	0.16
CO_2	—	—	—	—	0.07	0.03	0.06	—	—	—
S	—	—	—	—	—	—	—	—	—	—
Total	99.42	95.16	95.72	98.24	98.10	99.59	99.64	99.61	99.70	99.59
MgO/MgO +FeO	0.49	0.44	0.48	0.30	0.76	0.35	0.40	0.55	0.53	0.23

Normative minerals

	1	2	3	4	5	6	7	8	9	10
Q	4.2	—	6.6	0.5	—	6.1	—	15.0	8.6	9.4
C	—	—	—	—	—	—	—	4.8	—	—
Or	1.5	3.1	1.2	1.0	0.4	0.5	0.7	12.8	2.5	—
Ab	16.7	13.3	15.9	22.4	6.9	24.1	30.3	11.0	25.3	52.4
An	32.4	37.0	32.0	29.9	16.8	27.2	22.1	18.1	26.4	13.5
Ne	—	—	—	—	—	—	—	—	—	—
Di	18.8	9.1	15.8	11.4	8.9	17.4	16.5	—	9.8	3.0
Hy	25.3	31.8	27.4	32.3	27.1	23.1	12.6	37.6	26.6	18.2
Ol	—	4.0	—	—	38.2	—	14.9	—	—	—
Il	1.0	1.7	1.0	2.6	0.73	1.4	2.4	0.6	0.6	2.6
Ap	0.15	—	—	—	0.3	0.2	0.3	0.1	0.1	0.8

[a] Norms calculated on conversion of all Fe_2O_3 to FeO, deletion of H_2O and CO_2 and normalizing.

(1) Augite-rich dike in Lower Pillow Lavas, Cyprus, Bear (1960) Table III, p. 58.
(2) Altered basalt in Lower Pillow Lavas, Cyprus, Moores and Vine (1971) Table II.
(3) Altered dike margin in Lower Pillow Lavas, Cyprus, Moores and Vine (1971) Table II. (4) Altered basaltic pillow margin in Lower Pillow Lavas, Moores and Vine (1971) Table II. (5) Ultrabasic Lava in Upper Pillows, Bear (1960) p. 11. (6) Basalt, Eastern Papua, Davies (1971) Table 2, p. 26. (7) Basalt, Eastern Papua, Davies (1971) Table 2, p. 26. (8) Spilite feeder dike in pillow lavas, Semail ophiolite, Oman, Glennie et al. (1974) Table 6.9. (9) Spilite pillow basalt, Semail ophiolite, Oman, Glennie et al. (1974) Table 6.9. (10) Spilite flow in pillow sequence, Semail Ophiolite, Oman, Glennie et al. (1974) Table 6.9. (11) Spilite, Cerro Alto,

Table 7 (continued)

11	12	13	14	15	16	17	18	19	20
50.70	52.00	52.10	65.40	68.04	47.98	47.31	50.80	52.65	53.15
14.80	13.50	14.80	14.30	12.09	13.60	13.38	15.54	15.45	16.11
3.90	1.70	1.50	2.90	3.81	—	10.02	4.28	5.92	3.10
7.50	6.80	6.30	3.60	3.21	10.25	—	5.89	3.98	7.90
5.30	9.80	7.00	1.20	1.97	6.50	7.87	7.06	4.96	4.46
6.40	7.30	6.10	3.30	3.41	9.14	9.93	3.94	5.66	3.70
4.70	3.70	4.40	5.40	5.04	2.86	2.46	4.99	5.10	3.85
0.10	0.54	0.80	0.52	—	1.73	0.79	0.98	0.78	2.24
2.20	3.20	3.90	2.10	1.89	4.40	6.46	3.38	2.53	3.49
1.80	0.50	0.72	0.65	0.46	2.63	1.35	0.57	0.64	1.17
0.01	—	0.08	0.12	0.05	—	—	0.08	0.09	0.30
0.17	0.14	0.13	0.07	0.10	—	—	0.41	0.13	0.21
0.27	0.02	2.70	—	—	—	—	0.70	0.69	—
—	—	—	—	—	—	—	—	—	—
97.85	99.20	100.53	99.56	100.07	99.09	99.57	98.62	98.58	99.68
0.32	0.54	0.48	0.16	0.23	0.39	0.47	0.42	0.35	0.29

11	12	13	14	15	16	17	18	19	20
—	—	1.6	19.8	25.2	—	—	—	—	0.2
—	—	2.1	—	—	—	—	1.0	—	1.4
0.6	3.3	4.9	3.2	—	10.8	5.1	6.1	4.8	13.8
41.7	32.7	38.6	47.0	43.6	25.2	22.6	44.5	45.2	34.0
20.0	19.4	13.1	13.6	10.6	20.2	25.1	15.4	17.8	17.1
—	—	—	—	—	—	—	—	—	—
9.6	14.8	—	2.1	5.4	22.8	23.2	—	5.4	—
13.2	15.5	31.6	12.8	14.1	—	9.7	9.7	13.3	30.5
10.5	13.2	—	—	—	15.5	11.5	20.3	10.4	—
3.6	1.0	1.5	1.3	0.9	5.3	2.7	1.1	1.3	2.3
0.03	—	0.2	0.3	0.1	—	—	0.2	0.2	0.7

California, Bailey and Blake (1974) Table 4, p. 643. (12) Basalt, Pope Creek, California, Bailey and Blake (1974) Table 4, p. 643. (13) Spilite, Elder Creek, California, Bailey and Blake (1974) Table 4, p. 643. (14) Quartz Keratophyre, Bradford Mt., California, Bailey and Blake (1974) Table 5, p. 644. (15) Quartz Keratophyre, Del Puerto, California, Bailey and Blake (1974) Table 5, p. 644. (16) Basalt average of 5 analyses, Taurus, Turkey, Rocci (1973) Table 1. (17) Basalt, Mamonia Complex, Cyprus, Rocci (1973) Table 1. (18) Olivine basalt, Point Sal, California, Hopson et al. (1975) Table 1. (19) Aphyric basalt, Point Sal, California, Hopson et al. (1975) Table 1. (20) Pillow lava, Ergani Mine, Turkey, Bamba (1974) Table 1, p. 302.

of the rock with only minor ferromagnesian minerals. In a general way, the pillow lavas exhibit a mineralogy typical of tholeiitic basalts and in turn are quite similar to the underlying diabase dikes which act as feeders for the lavas. The widespread alteration of the pillow lavas has inhibited careful studies of the primary igneous minerals. As a result, there is virtually no data on the mineralogic nature of the igneous minerals. Increasing interest in oceanic pillow basalts dredged and drilled from the ocean floors is beginning to provide new and definitive compositional data on these igneous minerals and it is hoped that continued comparison between oceanic basalts and pillow basalts from ophiolites will stimulate additional work.

6.4 Chemistry

Perhaps more chemical data are available on pillow lavas than any other rock type of the ophiolites. Petrologists have not been prudent in their selection of material that has been analyzed nor have they documented the primary minerals or their alteration products in their published analyses. Therefore, any attempt to compare the analytical data from ophiolite pillows with unaltered fresh volcanic rocks from different settings is fraught with difficulties. The reader can find an appreciation of this problem by first reading Miyashiro's (1973a) attempt to classify the Troodos Ophiolite as an island arc assemblage based on plotting SiO_2, TiO_2, FeO* (total iron as FeO) against the FeO*/MgO ratio. Miyashiro's paper provoked considerable discussion in opposition to his handling of the chemical data versus known geologic considerations (Hynes, 1975; Moores, 1975; Gass et al., 1975). It is the opinion of the author that manipulating chemical data is important to the evolution of petrogenesis, but should always be tempered with geologic evaluation of the situation. In this section numerous plots will be discussed, but it must be borne in mind that too little is known about the analyzed rocks to assume that they all represent unaltered igneous rocks.

The bulk of the analysed pillow lavas from ophiolites are tholeiitic commonly having both normative hypersthene and quartz (Table 7). In some instances, advanced differentiation produces quartz-rich keratophyres consisting predominately of quartz and plagioclase. This same differentiation trend is also recorded in the sheeted dikes and the upper parts of the gabbroic sequences. It must be emphasized here that there has been found a diversity of volcanic rock types within ophiolites and that each complex may show unique trends. For instance, at Troodos, the lower pillow lavas appear to be silica-rich and according to

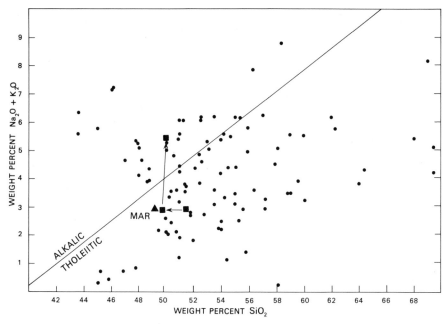

Fig. 25. $Na_2O + K_2O$ versus SiO_2 diagram of pillow lavas from ophiolites. MAR average mid-Atlantic Ridge basalt. All plotted in weight percent after subtracting total water and normalizing. Spilitic degradation of basalt shown by *squares* connected with *arrows*. (After Vallance, 1974)

Miyashiro (1973a, 1975a), are calc-alkalic or possibly strongly modified by post emplacement hydrothermal metamorphism (Gass and Smewing, 1973). In contrast, the upper pillows of the Troodos ophiolite are undersaturated olivine-bearing basalts associated with picrites. Mesorian et al. (1973) notes the presence of undersaturated nepheline-normative pillow lavas from some of the ophiolites from the eastern Mediterranean area.

A plot of $Na_2O + K_2O$ versus SiO_2 for a large number of analyses demonstrates the general tholeiitic nature of most of these lavas as opposed to lesser occurrences of undersaturated basalts (Fig. 25). It can also be demonstrated with this same diagram that the spilitic degradation of a tholeiitic basalt can effectively modify the original composition so that the altered rock assumes the chemistry of an alkalic basalt (Vallance, 1974). Plotting all of the available basalt analyses on Miyashiro diagrams (Fig. 26) provides another way to evaluate the nature of the parent magmas and their trends. The tholeiitic trend is characterized by iron enrichment in the early stages presumably because of crystallization and separation of olivines and pyroxenes with low

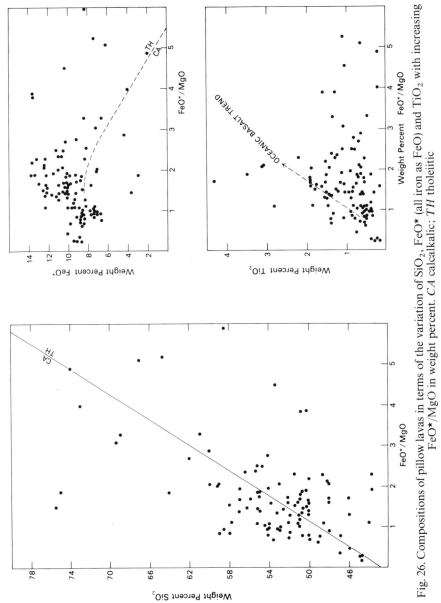

Fig. 26. Compositions of pillow lavas in terms of the variation of SiO_2, $FeO^\star$ (all iron as FeO) and TiO_2 with increasing $FeO^\star/MgO$ in weight percent. *CA* calcalkalic; *TH* tholeiitic

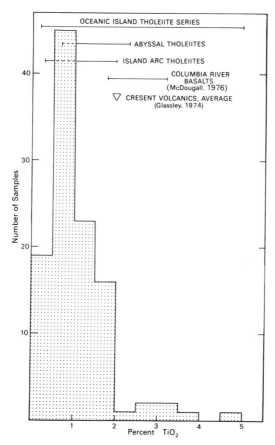

Fig. 27. Frequency distribution of TiO_2 (weight %) in pillow lavas from ophiolites compared to oceanic island tholeiites, abyssal tholeiites, and island arc tholeiites (Miyashiro, 1975a)

FeO*/MgO ratios. The plot of FeO*/MgO versus FeO* reveals that most of the pillow lavas show tholeiitic affinities.

However, the SiO_2 versus FeO*/MgO diagram seems to show that the analyses occupy both the tholeiitic and calc-alkalic regions of Miyashiro's diagram with the more differentiated rocks occupying both the calc-alkaline and tholeiitic fields. However, these more differentiated rocks are keratophyres consisting essentially of sodic plagioclase and quartz and are quite unlike the dacites and andesites of island arcs (Coleman and Peterman, 1975). This particular diagram (SiO_2 versus FeO*/MgO) does not allow a clear distinction between tholeiitic basalts from island arcs and abyssal tholeiites because of several shortcomings.

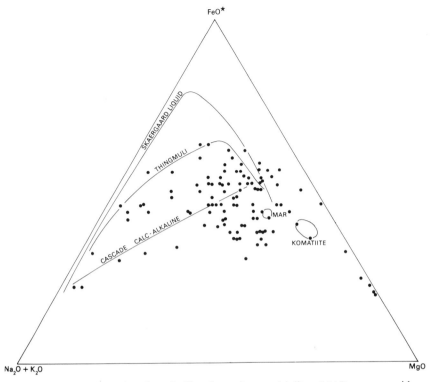

Fig. 28. AFM triangular plot of pillow lavas from ophiolites. MAR average mid-Atlantic ridge basalts. Archean komatiites are shown for comparison. All Fe converted to FeO* and normalized against the weight % of MgO and $Na_2O + K_2O$

First, it does not appear that there is enough difference between the bulk chemistry of island arc tholeiites and abyssal tholeiites to distinguish them (Jakes and Gill, 1970; Ewart and Bryan, 1972). Secondly, post-emplacement alteration of submarine basalts is so extensive that either loss or gain of silica could drastically change the SiO_2 versus FeO*/MgO plots (see the data from Vallance, 1974, plotted on Fig. 26).

Because TiO_2 does not appear to be affected as much as other elements, its trend on the TiO_2 versus FeO*/MgO plot (Fig. 27) may provide a better guide for comparison. The less differentiated pillow lavas show enrichment of TiO_2 during fractionation and follow the tholeiitic series, but the more differentiated lavas show a slight decrease in TiO_2 which could be characteristic of both the calc-alkaline and tholeiite series. Many granophyric parts of differentiated tholeiites are characteristically depleted in TiO_2 because of the crystallization of ilmenomagnetite in the middle stages of differentiation. A frequency

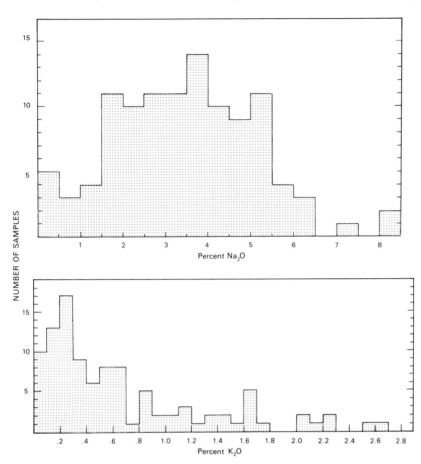

Fig. 29. Frequency distribution of Na$_2$O and K$_2$O (weight %) in pillow lavas from ophiolites

plot of TiO$_2$ from pillow lavas shows that the range (0.5–2.0 %) is very restricted and reflects the tendency for less iron and titanium enrichment in the ophiolite pillow lavas as compared to other tholeiite series (Fig. 27).

AFM plot of these same analyses shows an overall tholeiitic trend, but does not describe a strong iron enrichment (Fig. 28). Also, at least some of the analyses could be interpreted as being calc-alkalic with respect to both FeO* and Na$_2$O + K$_2$O. Some of the calc-alkaline tendencies could also be ascribed to hydrothermal alteration where Na$_2$O is added and FeO* removed (Bischoff and Dickson, 1975). Frequency distribution diagrams of K$_2$O and Na$_2$O in pillow lavas from ophiolites show a rather wide variation in comparison to TiO$_2$ (Fig. 29). The hydrothermal metasomatism accompanying the alteration of submarine pillow lavas must be a most important factor in changing the alkalis in these rocks. Vallance (1974) has found in progressive alteration of tholeiitic basalt that the alkalis are depleted in some stages and enriched in others.

Miyashiro (1975a) has attempted to classify ophiolites into three types in terms of their volcanic series and uses as the main basis of this classification variations of SiO_2, FeO^*, TiO_2 versus FeO^*/MgO. Class I would be ophiolitic complexes containing both calc-alkaline and tholeiitic volcanics. Class II are characterized by the presence of only tholeiitic series. Class III contain both tholeiitic and alkalic series. The advantage of this classification is that it attempts to identify the presence of igneous rock series within the volcanic suites of ophiolites and it is not based on the conceptual model that all ophiolites have to form at ancient mid-ocean ridges. The disadvantages of such a classification is that it deals with the rock type (volcanic) that is least abundant and usually most altered in most ophiolitic suites rather than considering it as an integral part of the total ophiolite assemblage. Furthermore, any classification of ophiolites based only on the chemical parameters of the volcanic suite without consideration of the fundamental structural or stratigraphic relationships of the total rock assemblage is destined to add confusion to a problem that still demands fundamental data prior to taxonomy (Moores, 1975; Gass et al., 1975). It now becomes apparent that formation of the pillow lavas within ophiolite sequences may have multiple origins or at least, parental magma of different starting compositions.

7. Geochemistry and Petrogenesis

7.1 Introduction

The formation of ophiolites will be explained in terms of their igneous development and even though the preceeding sections demonstrated that ophiolites occur as closely associated assemblages, their origins are polygenetic. Knowledge of the incompatible elements (Ba, K, Rb, Sr, Zr, U, Th, rare earths) and isotopes provides another means of tracing the evolution of igneous suites. Elemental abundances and isotopic ratios found in ophiolites combined with restrictions developed through experimental petrology can be used to distinguish igneous processes or eliminate models based on other parameters. Information on trace elements, isotopes, and petrogenesis of ophiolites is not abundant. The material to be presented in this section must, therefore, be utilized

Table 8. Average amounts of trace elements in peridotites, gabbros, dikes-lavas, plagiogranites from ophiolites in PPM

Metamorphic peridotite		Ultramafic[a]	Gabbro	Dikes and pillow lavas	Plagiogranite	MAR[b]	Tholeiites[c]
K	23	200	797	4200	3350	2158	5810
Rb	0.058	~ 1	1	5	2	—	30
Sr	0.38	~ 20	116	219	126	123	400
Ba	2	~ 0.4	~ 2	20	10	12	244
U	0.008	~ 0.02	—	—	—	—	—
Th	0.01	~ 0.06	—	—	—	—	—
Zr	5	30–45	~ 15	51	70	100	100
Co	114	110	86	63	7	41	40
Cu	14	~ 30	35	45	1	87	127
Ni	2280	~ 1500	640	215	10	123	85
Ti	48	300	1355	5102	3750	8300	9600
V	25	40	226	543	30	289	251
Pb	0.02	0.05	—	—	—	—	—
Cr	5000	2400	623	199	< 10	296	162

[a] Goles (1967).
[b] Melson et al. (1971), MAR = mid Atlantic Ridge basalts.
[c] Prinz (1967); Manson (1967).

with caution as the appearance of new data could seriously affect the conclusions that are based on such a small data base. As before, the ophiolite assemblage will be divided into specific groups: metamorphic peridotite, mafic and ultramafic cumulates, plagiogranites, and sheeted dikes-pillow lavas. It is presumed that each rock group is representative of a process than can be defined in terms of the magmatic evolution of the ophiolite assemblage and that the trace element abundance and isotopic ratios of each group will be controlled by identifiable processes.

7.2 Trace Elements

Establishing average values of trace elements for the ophiolite assemblage is a difficult task. Up to the present time, there does not exist complete analytical data on the trace elements or isotopes from any single ophiolite occurrence. The average values presented in Table 8 represent piecemeal data from various sources and from some rocks that may not be part of an ophiolite. Nonetheless, there emerges a consistency of data that provide a basis for discussion.

The metamorphic peridotites are obviously depleted in all the incompatible elements when compared to the overlying associated cumulates and extrusives. Goles (1967) has made a compilation of trace elements in all ultramafic rocks including those from ophiolites (Table 8) and it is clear that the metamorphic peridotites are also relatively depleted in the incompatible elements with respect to Goles' average. For instance, K, Rb, Sr, U and Zr in the metamorphic peridotites are one or two orders of magnitude less than that in the average ultramafic rock. It is also plain that the abundances of K, Th, and U in the metamorphic peridotites are all so low that calculations of heat production from these elements indicate that the metamorphic peridotites could not be an important component of either oceanic or continental mantle (Birch, 1965; Stueber and Murthy, 1966). No data exist on U or Th for rocks from ophiolite extrusives or cumulates, but K increases from 23 ppm in the metamorphic peridotite to 3350 to 4200 ppm in the plagiogranites and extrusives. These low K values are similar to the average Mid-Atlantic Ridge abyssal basalts (2158 ppm), but lower than that of the average given for all tholeiitic basalts (5810 ppm) (Table 8).

Ni, Co, and Cr are strongly enriched within the metamorphic peridotites and decrease by several orders of magnitude within the extrusive and cumulate rocks and are absent from the plagiogranites (Table 8, Fig. 30). Cr may reach similar values at the base of the cumulate section as are found in the underlying metamorphic peridotites since cumulate chromites are often concentrated here. Ni and Co abundances are directly related to the amounts of modal olivine and pyroxenes in the rock. V and Ti are both depleted in the metamorphic peridotites when compared to the cumulate and extrusive parts whose V and Ti are similar to the trends found in differentiating tholeiitic magmas (Table 8, Fig. 30). That is, the V and Ti become concentrated in the titanomagnetite crystallizing in the basalts and diabase with depletion of V in the plagiogranite (Carmichael et al., 1974, p. 480). Zr is depleted in the metamorphic peridotites (5 ppm) and shows a progressive enrichment from the cumulates (15 ppm) through the extrusives (51 ppm) to the plagiogranites (70 ppm) (Fig. 30). The trend for Cu in the ophiolites is interesting in that it is depleted both in the metamorphic peridotite and plagiogranite. None of the Cu abundance values within the various ophiolite parts is as high as those recorded for the average tholeiitic basalts (Table 8) suggesting either a primary magma low in Cu or possible postigneous removal of Cu as a result of hydrothermal alteration. Inclusion of continental tholeiites in averages will bias the average Cu values. As noted earlier, the widespread low-grade thermal metamorphism observed in the extrusive parts of ophiolites could lead to a depletion of Cu. Bischoff and Dickson (1975), using controlled experiments on seawater-basalt at 200° C, observed two or three orders of magnitude enrichment in Fe, Mn, Cu, and Ni in the reacted fluid over normal seawater. Additional quantitative data is required to evaluate the

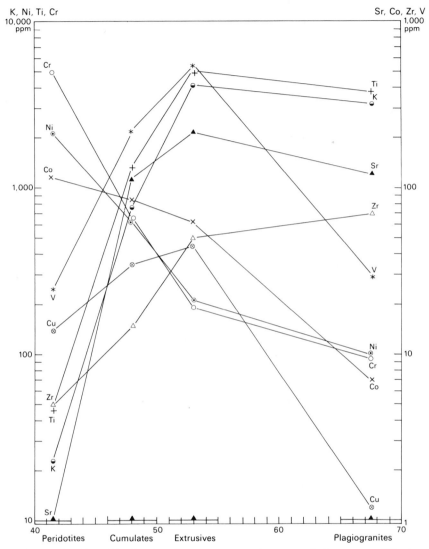

Fig. 30. Semilogarithmic plot of average trace element variation versus SiO_2 in metamorphic peridotites, cumulate rocks, extrusive rocks and plagiogranites in ophiolites

effect of such alteration with regard to the other trace elements in ophiolites, but it should be emphasized again that pervasive hydrothermal alteration could drastically alter the trace element abundances.

Rare earth elements (REE) can now be routinely determined by neutron activation or by mass spectrometry and provide additional

Table 9. Average rare earth contents for ophiolite sequences normalized to chondrites

No. of Samples	(5) Metamorphic peridotite		(4) Mafic cumulates		(4) Upper level gabbros		(2) Plagiogranite		(10) Lavas, dikes	
	Average	Normalized	Average	Normalized	Average	Normalized	Average	Normalized	Average	Normalized
Ba	1.9	0.52	2.1	0.57	6.0	1.6	27.1	7.4	42.3	11.6
La	0.068	0.23	0.065	0.22	0.443	1.48	3.08	10.3	2.8	9.3
Ce	0.072	0.09	0.316	0.38	1.68	2.0	8.95	10.6	5.74	6.8
Pr	0.006	0.05	—	—	—	—	—	—	1.75	14.5
Nd	0.026	0.04	0.384	0.66	1.60	2.76	8.69	14.98	5.56	9.6
Sm	0.008	0.04	0.173	0.82	0.644	3.07	3.40	16.2	1.96	9.3
Eu	0.0043	0.06	0.119	1.61	0.323	4.36	1.01	13.6	0.75	10
Gd	0.0125	0.04	0.297	0.93	1.38	4.31	5.43	17.0	3.03	9.5
Tb	0.0007	0.0014	—	—	—	—	—	—	0.47	1.0
Dy	0.0175	0.06	0.424	1.40	1.41	4.65	7.07	23	3.61	11.9
Ho	0.003	0.04	—	—	—	—	—	—	1.00	13.7
Er	0.0215	0.10	0.275	1.31	0.763	3.63	4.65	22.1	2.33	11.1
Tm	0.001	0.03	—	—	—	—	—	—	0.33	11.
Yb	0.016	0.094	0.306	1.8	0.74	4.35	4.59	27	2.17	12.8
Lu	0.004	0.129	0.053	1.7	0.126	4.1	0.85	27	0.312	10.1

information on the origin of ophiolites. Various authors have provided evidence that the REE are less liable to weathering and hydrothermal alteration than other elements and, therefore, their abundance patterns may provide clues to their igneous origins (Sugisaki et al., 1972; Montigny et al., 1973; Kay and Senechal, 1976). There are now available considerable data on the abundances of REE on various volcanic rocks and this provides a background with which to compare abundances from the various parts of the ophiolite sequence (Table 9). REE studies by Kay and Senechal (1976) on Troodos, Cyprus and Allegre et al. (1973) on Pindos, Greece are the only two complete studies available on ophiolite sequences. For the purposes of comparison, the absolute REE abundance in the samples is normalized by the corresponding element abundance in chondritic meteorites and plotted as a function of increasing atomic number (Haskin et al., 1966; Fig. 31). Averaging REE abundances from each part of the ophiolite sequence provides a basis to compare distributions with other igneous rocks (Fig. 31). The extrusive rocks have a pattern similar to that found for abyssal tholeiites in that the REE have low abundances when compared to most other basalts and nearly mimics the chondrite distribution. There is a slight depletion in Ce with positive anomalies in Ba and La (Fig. 31). The upper level gabbros follow the same broad pattern, but are depleted in all elements with a definite lowering of the light REE. A small Eu anomaly indicates its enrichment in cumulate plagioclase where it easily substitutes for Ca. The same Eu anomaly is even more pronounced in the mafic cumulate gabbros where the early fractionation of more calcic plagioclase is depleting the melt in Eu. The REE abundance pattern of the mafic cumulate gabbros falls well below the chondrite average for light REE but is nearly equal to the heavier rare earth abundances from chondrites. The plagiogranites in contrast to these trends show enrichment of nearly all REE when compared to other igneous members of the ophiolite. Particularly striking is the negative Eu anomaly which can be traced to the early abstraction of Eu by the calcic plagioclase in the cumulate gabbros and thus depleting the residual magma.

Figure 32 contrasts the REE patterns for plagiogranites, Precambrian tonalite, and granophyres. The depletion of heavy REE in the Precambrian tonalite is considered to represent deep melting in the mantle of eclogite or amphibolite. In contrast, the plagiogranite of the Troodos complex is enriched in heavy REE and has a prominent negative Eu anomaly similar to the REE pattern developed in granophyres produced by shallow differentiation of tholeiitic magmas. The REE patterns of the cumulate and extrusive parts of the ophiolite provide convincing evidence that these rocks are comagmatic and are

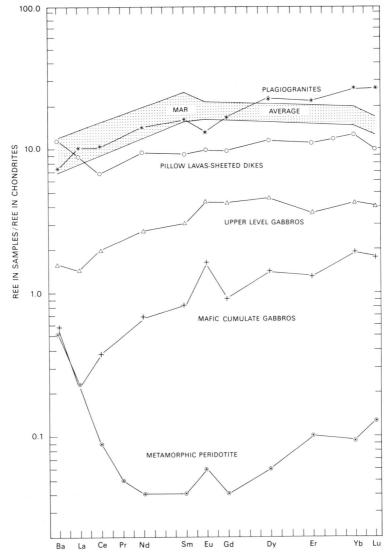

Fig. 31. Average rare earth element (REE) patterns for separate members of the ophiolite sequence REE normalized to standard chondritic abundance

related by the fractionating processes within a basaltic magma. Allegre et al. (1973) have calculated the REE abundances in the cumulate fraction and residual magma using partition coefficients for plagioclase, augite, and orthopyroxene. Assuming a cumulate assemblage of 60% plagioclase, 25% augite, 10% orthopyroxene and crystal fractionation

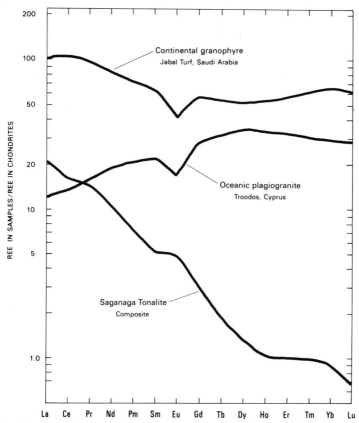

Fig. 32. Chrondrite normalized rare earth patterns for oceanic plagiogranite (Kay and Senechal, 1976), continental granophyre (D.G.Coles, 1974, unpubl.), and Precambrian tonalite (Arth and Hanson, 1972)

of 10%, 50% and 80% amounts of the original magma, they duplicate the observed average patterns found for the cumulate and extrusive parts of the ophiolite (Fig. 33).

The REE pattern of the metamorphic periodites cannot be related in any way to the unmistakable magmatic differentiation scheme of the overlying parts of the ophiolite. The metamorphic peridotites are depleted in nearly all REE by about two orders of magnitude (Fig. 31) and there is no possible way of deriving a partial melt of basaltic composition from these depleted metamorphic peridotites having a REE concentration similar to the overlying cumulates and extrusives. The REE data illustrates that these metamorphic peridotites must be residual fractions of an earlier partial melting from an undepleted primary mantle which could also have generated the basaltic oceanic crust.

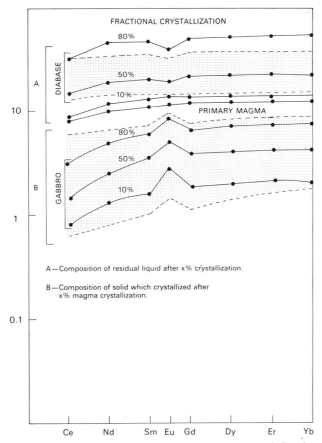

Fig. 33. Calculated REE distribution developed by crystal fractionation of a magma developed from partial fusion in the mantle. Percentages given on *curves* represent amount of solidified primary magma. *Shaded areas* represent spread of values determined on gabbros and dolerites from ophiolites. (After Allegre et al., 1973)

Allegre et al. (1973) have estimated the REE distribution for partial fusion of hypothetical primary mantle material (Fig. 34). Their calculations demonstrate that the REE depletion and distribution in the residual mantle is nearly identical to that observed in the metamorphic peridotites. However, up until now, none of the mantle materials studied contains high enough concentrations of the incompatible elements required by Allegre's calculations and so the nature of the primary mantle remains controversial.

Comparison of trace elements in ophiolites with those in oceanic basalts has reinforced the concept that the igneous rocks within the

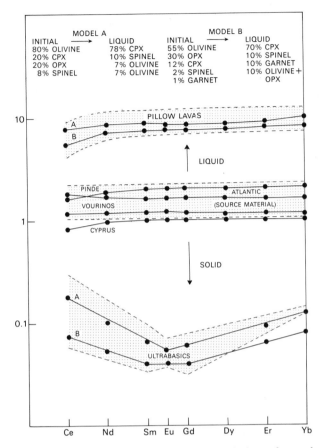

Fig. 34. Calculated REE distributions for partial fusion of mantle material. *Central shaded zone:* primary material; *various curves:* several types of primary mantle. *Upper shaded area:* partial melt; *lower shaded area:* residue of partial melting. *Single lines connecting dots:* calculated REE distribution assuming *Model A and B* with 10% partial melting. (After Allegre et al., 1973)

ophiolite sequence could have formed at a mid-ocean spreading axis. The REE patterns of the ophiolitic extrusives are remarkably similar to basalts dredged from the mid-ocean ridges (Montigny et al., 1973; Kay and Senechal, 1976; See also Fig. 31); however, it is not possible to readily distinguish the ophiolite REE patterns from basalts formed in small ocean basin or island arc tholeiites (Jakes and Gill, 1970). Some workers suggest that the abundance of Ti, Zr, Y may characterize basalt formed in different tectonic settings (Pearce and Cann, 1971, 1973; Pearce, 1975). The basic assumption is that these elements like the REE, are not mobile during weathering or low-grade hydrothermal alteration

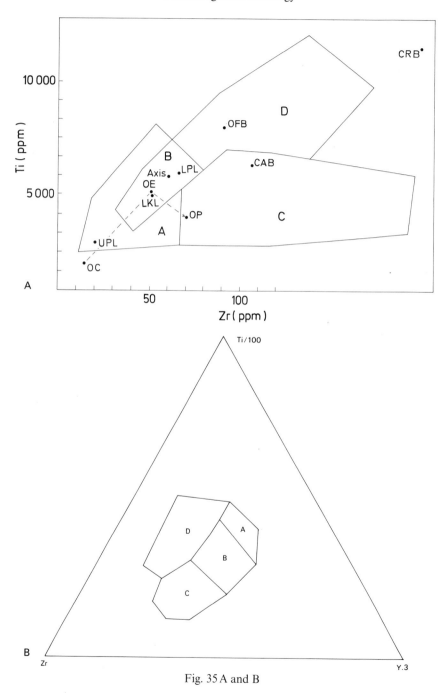

Fig. 35 A and B

Table 10. Average Ti, Zr, Y abundances for ophiolites and related rocks

Locality and source		PPM		
		Ti	Zr	Y
1. Lower pillow lavas, Troodos, Cyprus	(Pearce, 1975)	6175	66	28
2. Upper pillow, Troodos, Cyprus	(Pearce, 1975)	2518	20	15
3. Axis sequence, Troodos, Cyprus	(Smewing et al., 1975)	5900	61	28
4. Ocean floor basalts OFB	(Pearce, 1975)	8573	92	30
5. Island arc tholeiitis LKT	(Pearce, 1975)	4975	52	19
6. Calc-alkali basalts CAB	(Pearce, 1975)	6475	106	23
7. Ocean island basalts OIB	(Pearce, 1975)	18704	215	29
8. Columbia River basalt CRB	(McDougall, 1976)	13524	202	53

and could, therefore, fingerprint the nature of the basaltic magmas. Pearce has devised several plots to graphically display the distribution of $Ti-Zr-Y$ and $Ti-Zr$ (Fig. 35 A, B) and has delineated fields for ocean floor basalts, island arc basalts, and calc-alkalic basalts (Table 10). Application of these plots leads to some ambiguity in that there is considerable overlap between the various designated fields and as in the case with REE, it is not possible to clearly distinguish between island arc tholeiites and ocean floor basalts. Average values for Ti and Zr from various parts of an ophiolite sequence demonstrate that they vary according to degree of differentiation and describe a trend from field A through B and then into C. Therefore, to apply this kind of plot, it is essential to use the least differentiated extrusive parts of the ophiolite sequence in order to provide a valid comparison with other basaltic types.

7.3 Strontium Isotopes

Application of strontium isotopes towards the study of ophiolites has been approached in two separate ways. The first data was obtained on the metamorphic peridotites with the idea that the data could lead to a better understanding of the upper mantle (Roe, 1964; Stueber and

Fig. 35. (A) Ti–Zr plot of Pearce and Cann (1973); ocean floor basalts in fields B and D, calc-alkalic basalts in fields B and C, island-arc basalts in field A and B. Average values are plotted for: LKL: island arc tholeiites; OFB: ocean floor basalts; CAB: calc-alkalic basalts; CRB: Picture Gorge member of the Columbia River basalts; Axis: Troodos ophiolite, Cyprus axis sequence; OC: ophiolite cumulates; OE: ophiolite extrusives; OP: ophiolite plagiogranite; UPL: Troodos ophiolite, upper pillow lavas; LPL: lower pillow lavas. (B) Ti–Zr–Y plot of Pearce and Cann (1973); ocean floor basalts in field B; island arc basalts in field A and B, calc-alkalic basalts in field B and C, within plate basalts in field D

Table 11. Strontium isotope and K, Rb, Sr, abundances in ophiolites PPM

	K	Rb	Sr	K/Rb	Rb/Sr	Sr^{87}/Sr^{86}
Pillow lavas and sheeted dikes						
1. Basalt flow, Troodos, Cyprus Peterman et al. (1971)	9296	7.7	104	1207	0.074	0.7053
2. Sheeted diabase, Troodos, Cyprus Peterman et al. (1971)	2241	2.2	98	1018	0.022	0.7056
3. Sheeted diabase, Troodos, Cyprus Peterman et al. (1971)	4731	3.1	107	1526	0.029	0.7057
4. Sheeted diabase, Troodos, Cyprus Peterman et al. (1971)	3403	3.4	134	1000	0.025	0.7057
5. Basalt, upper pillows, Cyprus Peterman et al. (1971)	3569	10.6	163	336	0.065	0.7048
Plagiogranites						
1. Plagiogranite, Troodos, Cyprus Peterman et al. (1971) Cy 13	2822	1.9	134	1485	0.014	0.7045
2. Keratophyre, Troodos, Cyprus Coleman and Peterman (1975) Cy 55 B	2739	1.6	109	1712	0.105	0.7055
3. Plagiogranite, Troodos, Cyprus Coleman and Peterman (1975) Cy 55 C	581	< 1	135	7581	<0.007	0.7057
4. Plagiogranite, Troodos, Cyprus Coleman and Peterman (1975) Cy 52	2490	3.4	148	732	0.023	0.7053
5. Plagiogranite, Troodos, Cyprus Coleman and Peterman (1975) Cy 55 A	5312	3.4	114	1562	0.030	0.7059
Cumulate rocks						
1. Olivine gabbro, Troodos, Cyprus Peterman et al. (1971) Cy 10	700	1.3	41	538	0.032	0.7049
2. Clinopyroxene gabbro, Troodos, Cyprus Peterman et al. (1971) Cy 4	706	—	86	—	<0.01	0.7040
3. Uralite gabbro, Troodos, Cyprus Peterman et al. (1971) Cy 3	2158	4.3	83	501	0.052	0.7040
4. Olivine gabbro, Troodos, Cyprus Peterman et al. (1971) Cy 5	623	—	102	—	<0.01	0.7048
5. Uralite gabbro, Troodos, Cyprus Coleman and Peterman (1975) Cy 54	1909	1.1	42.9	1740	0.026	0.7065

Sample						
6. Plagioclase from # 5 Coleman and Peterman (1975) Cy 54	2822	1.3	98.7	2170	0.013	0.7057
7. Two pyroxene gabbro, Troodos, Cyprus Coleman and Peterman (1975) Cy 50	166	<1	31.7	>166	<0.03	0.7048
8. Plagioclase from # 7 Coleman and Peterman (1975) Cy 50	996	1.4	67.1	711	0.02	0.7043
9. Uralite gabbro, Troodos, Cyprus Coleman and Peterman (1975) Cy 50 A	166	<1	116	>166	<0.009	0.7049
10. Anorthosite, Troodos, Cyprus Coleman and Peterman (1975) Cy 53	249	<1	137	>249	<0.007	0.7040
11. Gabbro, Yate, New Caledonia. Coleman and Peterman (1975) Cy 55 d	1100	<1	141	>1100		0.7033
12. Gabbro, Pirouge River, New Caledonia, Z. Peterman, unpubl. data 76-Nc-62	124	<1	204	>124		0.7035
13. Gabbro, Bolu Creek, Papua-New Guinea. Z. Peterman unpubl. data 82-NC-62 A	53	<1	49	>64		0.7031

Z. Peterman unpubl. data 0458

Metamorphic peridotites

Sample						
1. Dunite, Dun Mt. New Zealand Stueber and Murthy (1966)	23.9	0.111	4.39	215	0.025	0.7091
2. Dunite, Papua Stueber and Murthy (1966)	61.2	0.302	6.28	203	0.048	0.7078
3. Peridotite, Mt. Albert, Quebec Stueber and Murthy (1966)	59.3	0.158	5.52	375	0.029	0.7109
4. Dunite, Addie-Webster, North Carolina Stueber and Murthy (1966)	15.0	0.077	2.99	195	0.026	0.7156
5. Peridotite, Tinaquillo, Venezuela Stueber and Murthy (1966)	25.8	0.093	3.89	277	0.024	0.7084
6. Peridotite, New Caledonia	—	8.35	13.7	—	0.609	0.7066
7. Dunite, New Caledonia	—	0.073	0.523	—	0.15	0.7079
8. Dunite, New Caledonia Roe (1964)	—	0.147	0.499	—	0.295	0.7127
9. Peridotite, Black Lake, Quebec Roe (1964)	—	0.380	0.787	—	0.484	0.7149
10. Peridotite, Black Lake, Quebec Roe (1964)	—	0.307	3.60	—	0.089	0.7096
11. Clinopyroxene in Pyroxenite, Canyon Mt. Lanphere (1973)	—	—	—	—	—	0.708

Murthy, 1966; Stueber, 1969). Many of these samples are poorly documented and, therefore, it is difficult to evaluate the results (Faure and Powell, 1972, p. 72). Another approach was to determine the strontium evolution in ophiolite extrusives and compare this with modern oceanic basalts (Bonatti et al., 1970; Montigny et al., 1970; Peterman et al., 1971; Coleman and Peterman, 1975). As a result of these two divergent approaches, there is not yet available a published report on a single ophiolite complex where complete Sr isotope data are available on both the peridotites and overlying extrusive-cumulate sequence. This discussion will, therefore, have to combine the available data from the various sources in order to allow a partial synthesis relative to ophiolites (Table 11).

The metamorphic peridotites have extremely low Rb (~ 0.18 ppm) and Sr (~ 3 ppm) and, therefore, the analytical problem in determining the Sr^{87}/Sr^{86} ratios is very difficult. Furthermore, contamination during serpentinization and metamorphism by outside fluids could very easily affect the very low amounts present in these rocks (Faure and Powell, 1972). The range in Sr^{87}/Sr^{86} for the metamorphic peridotites is from 0.7078 to 0.7156 and is surprisingly high considering its probable source within the mantle (Table 11). The cumulate gabbros contain on the average ~ 1.9 ppm Rb and ~ 80 ppm Sr, whereas, the sheeted dikes-pillow lavas average 5 ppm Rb and 121 ppm Sr both considerably higher than the underlying metamorphic peridotite. The Sr^{87}/Sr^{86} ratios for the cumulates and extrusives range from 0.7040 to 0.7065 similar to the oceanic basalts, but somewhat higher than the average oceanic ridge basalt. Plagiogranites within the ophiolites have similar Sr^{87}/Sr^{86} ratios (0.7045–0.7059), an expression of their comagmatic nature with the cumulates and extrusives.

The Sr^{87}/Sr^{86} ratio in the metamorphic peridotites is higher than in the cumulates and extrusives of the ophiolites or any other volcanic rock from the deep oceans (Fig. 36). Furthermore, these high values are actually higher than the Sr^{87}/Sr^{86} initial ratios of continental basalts and most other silica-rich igneous intrusions. Assuming that the present analytical data available reflect the true initial ratios of Sr^{87}/Sr^{86} in the metamorphic peridotites, there appears to be no genetic relation between them and the associated cumulates and extrusives of the upper parts of the ophiolite sequence. The strontium isotopes provide another independent line of evidence that shows the polygenetic nature of the ophiolite assemblage. The metamorphic peridotites have an unmistakable history that clearly separates them from the cumulates and extrusives. If the metamorphic peridotites represent a residue of partial melting and it is clear that their depleted nature indicates this, then the explanation of the strontium isotope ratios require that this event took place early in the history

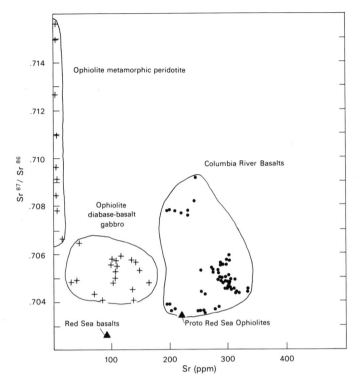

Fig. 36. Sr^{87}/Sr^{86} versus Sr for ophiolites *(crosses)* and other rocks. Data for Columbia River basalts from McDougall (1976) and Red Sea basalts-ophiolite from Coleman et al. (1975b).

of the mantle, at least one billion years ago (Faure and Powell, 1972) (Fig. 37). Such restrictions based on trace elements and isotopes require that the source for the magma that forms the cumulates and extrusive parts of the ophiolite cannot be the metamorphic peridotite that forms their base. The general trace element and isotopic in compatibility between the ultramafic xenoliths and the enclosing volcanic rocks indicates that there are large areas in the mantle that may have a history similar to the metamorphic peridotites from ophiolites. The metamorphic peridotite that forms the base of ancient ophiolites could, therefore, represent similar mantle residue from a much earlier melting event within the mantle.

7.4 Petrogenesis

The ophiolite assemblage consists of three main parts: (1) metamorphic peridotite; (2) cumulate sequence; (3) sheeted dikes and pillow lavas. Each of these parts has formed in response to processes that are

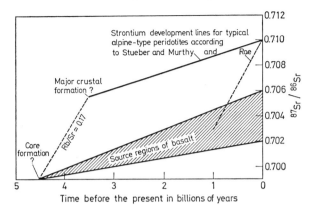

Fig. 37. The Sr^{87}/Sr^{86} ratio development diagram for alpine-type ultramafic rocks according to the model of Roe (1964) and the model of Stueber and Murthy (1966). The diagram is modified from those shown by Stueber and Murthy (1966) and Hurley (1967). Also included in *shaded zone* that shows the growth of Sr^{87}/Sr^{86} ratio in the source regions of oceanic basalts with time. According to both models the alpine-type ultramafic rocks have been derived from a residual zone in the mantle with little if any accompanying change in their Rb/Sr and Sr^{87}/Sr^{86} ratios. The very low strontium contents of alpine-type ultramafic rocks, however, render them extremely susceptible to contamination with radiogenic strontium, and if they have been contaminated, their present Rb/Sr and Sr^{87}/Sr^{86} ratios are not directly related to those of their deep source regions. (After Faure and Powell, 1972, Fig. VII. 1)

obviously polygenetic, but the apparent stratigraphic sequence suggests that each process may have been penecontemporaneous with the other except for the earlier melting of the metamorphic peridotite.

Let us look at each part of the assemblage and determine its relationship with the other parts based on the information developed in the previous sections. The metamorphic peridotites characteristically have uniform mineral modes and except for the chromite segregations exhibit metamorphic fabrics that have formed in response to subsolidus deformation under mantle conditions. The extremely low levels of incompatible elements and the unusually high Sr^{87}/Sr^{86} ratios indicate that the metamorphic peridotites represent a residual fraction of the mantle that may have undergone partial melting at least 1 b.y. ago. The metamorphic peridotites form the bottom parts of ophiolites and in some cases are the most extensively exposed rocks or in other cases may be completely tectonically separated from the other parts of the ophiolite sequence. It therefore seems unlikely that the metamorphic peridotites can be considered contemporaneous residual mantle that remained after producing the mafic magma for the overlying cumulate and extrusive rocks. Rather, these depleted metamorphic hazburgites represent much older parts of the mantle that have moved tectonically upward to form the base upon which the upper parts of the ophiolite assemblage develop. Detection of the transition zone between these underlying older metamorphie peridotites and the overlying ultramafic cumulate rocks is difficult as the mineralogy and field appearance of these two parts are very similar and often a common deformational fabric is superimposed on both parts. However, the composition of the olivine, orthopyroxene, clinopyroxene, and chromite in the metamorphic peridotites is distinctly different and the Sr^{87}/Sr^{86} ratios are much higher (see Table 11) than the cumulate peridotites. Furthermore, apophyses of the cumulate section often penetrate downward into the metamorphic peridotite and crosscut the metamorphic fabric, demonstrating their later development.

The cumulate sections of the ophiolite sequences range from dunites through gabbros to plagiogranites and indicates that large amounts of mafic magma underwent crystal fractionation. Generally the cumulate sections are nonuniform and show sharp changes laterally and vertically that cannot be reconciled with the concept that the layering results from the static crystallization of one large batch of magma as visualized for the Skaergaard or Bushveld layered intrusions. Even though cyclic units have been reported from Vourinous, there has not yet been established a clear picture of the possible sequential events that lead to the development of layering in these ophiolite cumulates. The crystallization sequence in the cumulates follows the low-pressure differentiation trend of tholeiitic magmas beginning with olivine and chromite followed by olivine and clinopyroxene, with an early appearance of interstitial plagioclase. The ultramafic parts of the cumulate section are usually very restricted when compared to the thicknesses of the underlying metamorphic peridotites. In the upper parts of the cumulate section, plagioclase is the dominant phase with variable amounts of clinopyroxene, orthopyroxene, and olivine, with brown hornblende often an important phase. An interesting aspect of most ophiolite cumulate sections is the low amount of iron oxides and lack of iron enrichment of the silicates from the base upwards. The lack of iron in the cumulate section must signify a parental magma depleted in iron or an enrichment of iron in a residual liquid that never crystallized. Plagiogranites representing the residual differentiated liquids of the cumulate sequence are not enriched in iron, but actually show a depletion.

Determination of the parental magma for the cumulate parts of ophiolites is difficult because the cumulate sections do not represent a closed differentiation system and "chilled" border facies are not available. However, the constructional sheeted dikes and pillow lavas capping the cumulate gabbros may contain "chilled" fractions of the primary liquid that is parent to the underlying cumulates. This requires that the cumulates and the overlying extrusives are penecontemporaneous. The sheeted dikes exhibit chilled margins against one another and in the upper sections of the sheeted dikes, they unmistakably act as feeders for the pillow lavas whose submarine extrusion is manifested by their shape and interlayered pelagic sediments. The time relationships between the upper parts of the cumulate sequence and the sheeted dikes is however not clear from presently known field relationships. Fine-grained diabase dikes with chilled margins extend downward into coarser-grained layered gabbro and invasion of sheeted dikes through both the metamorphic peridotites and cumulate gabbros in Hatay, Turkey (Parrot, 1973) indicate that some sheeted dikes are derived from a later magma than that which produced the associated underlying gabbros. In other instances, it appears that the layered gabbros have been emplaced after the sheeted dikes and pillows formed and that these extrusives are a preexisting roof above the magma chamber. The actual igneous time sequence within the constructive parts of the ophiolite assemblage is variable and each occurrence may have its own unique intrusive sequence. Also, it is quite likely that different levels of crystal fractionation and intrusion could produce the conflicting intrusive sequences. Nonetheless, comparison of the chemical parameters within these cumulate and extrusive parts of the ophiolite assemblage provide unmistakable evidence of comagmatic evolution. The best evidence for the cogenetic nature of these rocks is based on the initial Sr^{87}/Sr^{86} ratios and REE distributions Figs. 31 and 36). The similarity of the Sr^{87}/Sr^{86} ratios between cumulate and extrusive rocks, as well as the REE partitioning produced by the cumulate processes, define a pattern of magmatic evolution from a similar parent magma.

The parent magma for the constructional parts of the ophiolite assemblage appears to be most closely related to the subalkaline abyssal oceanic tholeiites that are so common in present-day oceanic ridges. The low abundance of the dispersed elements, tholeiitic nature, REE distribution, and extrusion in deep oceanic environments is strongly suggestive that ophiolites represent fragments of ancient oceanic crust. Formation of such a magma is visualized as a partial melting of mantle material where the source material may have undergone earlier partial melting, as well as depletion of dispersed elements. Sr^{87}/Sr^{86} values for extrusive ophiolite are higher (0.7057–0.7040) than those from modern abyssal tholeiites (0.7020–0.7035) and suggest that at least some ophiolite magmas were derived from mantle depleted in Rb and incompatible elements earlier than the source of present-day basalts from spreading ridges. The significantly higher Sr^{87}/Sr^{86} values (0.7015–0.7064) of the underlying metamorphic peridotites precludes these rocks as being the residue of partial melting or cumulates to a parental magma contemporaneous with the overlying constructive ophiolite. If it is assumed that the parent magma for ophiolites is the same as abyssal tholeiites, then high-level differentiation of this magma prior to its eruption as dikes or pillow lavas would produce a much more evolved basalt than is characteristic of ophiolites. Differentiated basalts are not uncommon in ophiolites and so at least some differentiation of the primary magma must have taken place.

The occurrence of extrusive picrites (komatiites) in some extrusive parts of Paleozoic ophiolites (Gale, 1973) indicates that the primary liquid may have been more mafic and that the cumulate parts of the ophiolite could represent early crystal fractionation of a picritic magma. Explanations for komatiites and extrusive ultramafics (spinifex rocks) in the Archean demand much higher heat flow within the upper parts of the mantle than in the Phanerozoic (Brooks and Hart, 1974; Green et al., 1975); however, it appears possible that deep melting in the mantle could produce a picritic melt even during the Phanerozoic (O'Hara, 1968). This picritic melt could be modified by fractionation of olivine + chromite at high levels such as we see at the base of the ophiolite section and evolve towards the subalkaline abyssal basalt and ultimately to plagiogranite. It is possible to estimate liquid lines of descent from cumulate rocks given adequate samples and knowledge of the volumes of the various units; however, up until this time, such information is not available for cumulate sections from ophiolites. Therefore, the question of what is the primary magma for various ophiolite assemblages remains moot. Calculations based on average compositions within the cumulate parts of ophiolite sequence requires that the primary magma have a fairly low total iron content. The average abyssal tholeiite contains too much iron to be the parent for the cumulates, and if a picrite primary magma is selected, it is not possible to develop an abyssal tholeiite from such a magma by crystal fractionation using the cumulate assemblages found in ophiolites as a guide to fractionation. It is conceivable that the layered sequences of ophiolites represent trapped liquids that were unable to penetrate an overlying extrusive carapace. Walker (1975) has shown that the rise of basic magma in the Iceland dike swarms is dependent on density of the rising magma in contrast to the rocks it invades. The magmas that give rise to the cumulate parts of ophiolites may represent a magma trapped by an overlying denser carapace. In other situations where ascent of the magma is more rapid during periods of distension, the magma will reach the surface to produce extrusions.

The geologic and petrologic evidence assembled here and by other authors points towards an oceanic realm as the place of formation for ophiolites. Steinmann and other European geologists first pointed out that the nearly universal association of deep water pelagic sediments and pillow structures imposes definite restrictions on the geologic setting where ophiolites could form. The lack of intrusive contacts with surrounding rocks and the common association of mélanges later led to the general concept that most ophiolites were allochthonous slabs. It is now generally accepted that the ophiolite assemblages represent fragments of oceanic crust and, therefore, the igneous processes displayed within exposed ophiolites may be the same as those that give rise to the formation of oceanic crust at modern spreading ridges. Most modern authors who have studied ancient ophiolites have been impressed with the similarity of the rock types and gross structures of ophiolites with those of oceanic crust. There are, however, other workers who are convinced that the ophiolites represent processes characteristic of volcanic island arcs, geosynclinal submarine extrusions, or small ocean basis. Restrictions developed by petrologic studies by no means exclude those various possibilities and, in fact, in most cases by careful selection of data, it is possible to develop a plausible explanation for ophiolite generation in any one of those diverse geologic settings.

In all of the possible geologic settings for forming ophiolites, it is unmistakable that igneous processes leading to their development were either at the edge of continental crust or within the oceanic crust. No ophiolites have been related to igneous events characterizing stable platform areas or Precambrian shields. The extrusive parts of the ophiolites can be related to tholeiitic parent magmas and the underlying cumulates represent crystal fractionation of these same melts. Some authors have shown a calc-alkaline trend for the extrusives and suggest that this trend is typical of island arcs forming in oceanic areas. However, complete ophiolite sequences have as yet to be discovered in such island arcs. Radial dike distribution, isolated volcanoes, and thick pyroclastic sedimentary sections so typical of island arc assemblages are not present within well documented ophiolite sequences. It is entirely possible that some oceanic island arcs have oceanic crust as their basement, as suggested for Eua in Tonga (Ewart and Bryan, 1972). It has also been suggested that the concentrically zoned dunite–peridotite associations are exposed roots of island arc volcanoes and, if so, these ultramafic rocks can easily be distinguished from the metamorphic peridotites and cumulate ultramafic parts of the ophiolites.

Development of new oceanic crust at mid-ocean spreading centers has been documented at numerous sites. Present studies by DSDP and FAMOUS projects, as well as indirect geophysical measurements of the oceanic crust, provide evidence that suggests the oceanic crust has a stratigraphy similar to that found in ophiolites. Dredge and drill samples of the igneous rocks of the deep oceans all have their counterparts in the ophiolites, including metamorphic peridotites and cumulate rocks. Numerous models of mid-ocean spreading tied to observed petrologic relationships, have been published and they bear a startling similarity to the ophiolite sequences. Mainly because all these models draw heavily on spatial relationships observed for on-land ophiolites.

There are numerous aspects of known ophiolites which violate the mid-ocean spreading models and so some authors have suggested marginal basins or small ocean basins as a locale for generating igneous sequences similar to ophiolites. The only actualistic model for the formation of oceanic crust is present-day spreading at mid-ocean ridges. Other methods of forming new oceanic crust must relay on indirect evidence developed through petrologic reasoning or structural evolution of the specific area. It is apparent to me, however, that we must consider all the various possibilities listed above as likely situations where ophiolites can be generated. Petrologic considerations should not be the only means by which a particular evolutionary history of each separate ophiolite is established. Structural setting, as well as the evidence provided by the associated sediments, may provide the more definitive evidence towards elucidating each ophiolite's origin.

Part IV. Metamorphic Petrology

1. Introduction

Modification of the ophiolite assemblage by metamorphic processes that postdate the primary igneous origin are common. These metamorphic processes will be divided into two broad categories for the purposes of this discussion so that the reader can evaluate their relative importance.

1. Internal metamorphism of the ophiolites will include those metamorphic events that modify only the ophiolite mineral assemblages, such as serpentinization and thermal metamorphism.

2. External metamorphism will include metamorphic events affecting associated country rocks as well as the ophiolites. This metamorphism is related to the tectonic history of emplacement or later orogenesis that the ophiolites have been through. In the case of serpentinization, both external and internal metamorphism are interrelated.

The assumption will be made here that the ophiolites have formed either at a spreading ridge or equivalent situation where high heat flow obtained for substantial periods after the constructional parts of the ophiolite sequence developed. Furthermore, it will also be assumed that most ophiolites have been tectonically transported and have been incorporated into the orogens of continental margins. Thus, we have the potential for early thermal metamorphism in the oceanic realm, as well as later dynamo-thermal metamorphism related to the tectonic history of the ophiolite in the orogen. It is, therefore, conceivable that at least some of the older ophiolites may have undergone polymetamorphism resulting in nearly complete obliteration of their primary textures, as well as drastic changes in their chemistry (Table 12).

It must also be kept in mind that the products of metamorphism will have a great mineralogic diversity because we have within ophiolites a bulk compositional variation from peridotite to basalt with even small amounts of plagiogranite present.

Table 12. Scheme of metamorphism for ophiolites

Ophiolite protolith	Oceanic hydrothermal	Subduction	Obduction	Regional
Basalt	Zeolite (spilites) →	Zeolite →	No metamorphism	Zeolite (spilites) →
Diabase	Greenschist →	Prehnite-Pumpellyite →		Prehnite-Pumpellyite →
Gabbro	Amphibolite	Blueschist (Type C eclogite)		Amphibolite
Peridotite	Local serpentinization / Rodingites / Transform fault, mylonites, and cataclasites	Serpentinization ↘ ↗ Blueschist → Antigorite / Zeolite → Lizardite-Chrysotile / Rodingites	*Hot slab*: Dynamothermal aureole. Amphibolite Mylonite at base. *Cold slab*: Mélange Serpentinization Tectonic Overpressure? Rodingites	*Prograde metamorphism of serpentine*: Chrysotile → Antigorite → Olivine+Antigorite → Olivine+Talc → Olivine+Enstatite ; Rodingites
Conditions of metamorphism	Steep thermal gradients. No deformation. Minor metasomatism	Low thermal gradients. Deformation. Mélanges	Very steep thermal gradients / Very low thermal gradients. Deformation and metamorphism at base of ophiolite slab	Steep regional thermal gradients. Polymetamorphism. Sub-solidus deformation of peridotites

2. Internal Metamorphism

2.1 Serpentinization

The peridotites within ophiolites are usually serpentinized to some degree and it is rare to find peridotites that contain completely unaltered olivine, orthopyroxene, or clinopyroxene (Table 12). The process of changing a peridotite to serpentine involves mainly a hydration reaction between water and the primary igneous minerals. Mineralogically, serpentinites consist predominately of lizardite, clino-chrysotile, and antigorite with minor amounts of brucite, talc, magnetite, and carbonate. Chrysotile is characteristically fibrous with a tube-like structure. Lizardite and antigorite crystallize as plates and are not known to assume the fibrous habit of chrysotile. Present knowledge regarding these serpentine minerals indicates they are not isochemical, hence cannot be considered polymorphs and, therefore, the presence of any one species alone cannot specify a particular pressure-temperature condition (Coleman, 1971 b).

Synthesis of antigorite is difficult, but nonreversible breakdown reactions and D.T.A. analyses indicate that antigorite may be stable at higher temperatures than lizardite and chrysotile (Faust and Fahey, 1961; Yoder, 1967; Iishi and Saito, 1973). Additional evidence is the predominance of antigorite within serpentinites occurring in high-grade metamorphic terrains (Chidester, 1962, p. 73; Trommsdorff and Evans, 1972, 1974). The occurrence of antigorite with chrysotile and lizardite is common. However, where I have observed the assemblage antigorite, lizardite, and chrysotile, textural evidence suggested replacement of lizardite and chrysotile by antigorite or replacement of antigorite by lizardite and chrysotile. The serpentine minerals may have over-lapping fields of stability that are controlled by composition (amounts of Fe and Al), oxygen fugacity (Fe^{2+}–Fe^{3+} content), and activity of H_2O. Lizardite and chrysotile can form at temperatures less than $350°$ C down to ambient temperatures (Barnes and O'Neil, 1969), whereas antigorite is apparently stable up to temperatures slightly above $500°$ C. Recognition of serpentine species in ophiolite assemblages, therefore, can be of value in establishing P–T conditions during serpentinization.

During serpentinization of dunite, harzburgite, or lherzolite, assuming that only water in introduced, the possible mineral species that can be developed at these bulk compositions are limited as shown by Figure 38. In the case of small ultramafic bodies being serpentinized or undergoing regional high-grade metamorphism, it is not unusual to have CO_2 or silica invade the ultramafic body at its borders to produce talc, chlorite,

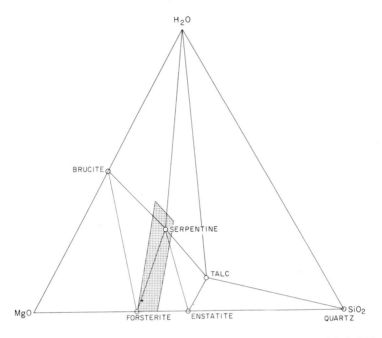

Fig. 38. Triangular diagram illustrating mineral species in the system MgO–SiO$_2$–H$_2$O with *superposed field* of compositional range for serpentinites and their protolith peridotites from ophiolites

and carbonates (Chidester, 1962, 1969; Jahns, 1967), within the serpentinite.

The assemblage lizardite + clinochrysotile + brucite + magnetite appears to be the most common one developed in the peridotites from ophiolites. Where these same ultramafic rocks are found in terranes that have undergone external high-grade regional metamorphism, antigorite is the stable serpentine mineral. Progressive metamorphism of peridotites from the Pennine region of the central Alps provides a consistent picture of the assemblages developed in the system MgO – SiO$_2$ – H$_2$O (Trommsdorff and Evans, 1974) (Fig. 39). If the experimental information can be used as a guide to pressure-temperature conditions, it would seem that most of the serpentinites derived from ophiolite peridotites probably formed in the temperature range 100°–300° C with only minor amounts forming at ambient temperatures (Wenner and Taylor, 1971). The antigorite serpentinites must represent temperatures in excess of 300° C and perhaps as high as 550° C and could reflect higher grades of regional metamorphism.

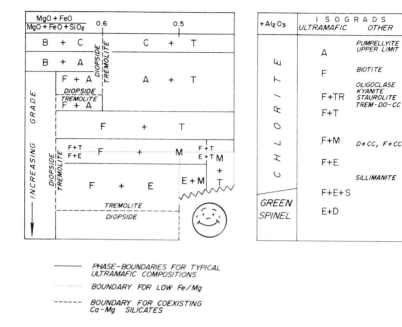

Fig. 39. Summary of possible mineral assemblages within progressively meta-morphosed peridotite-serpentinite (after Trommsdorff and Evans, 1974). *A* anti-gorite; *B* brucite; *C* chrysotile, lizardite; *CC* calcite; *D* diopside; *DO* dolomite; *E* enstatite; *F* olivine; *M* magnesiocummingtonite/anthophyllite; *S* spinel; *T* talc; *TR* tremolite

Assuming that serpentinization involves only olivine and water hypothetical reactions representing the most probable chemical changes can be written for serpentinization of peridotites:

$$2\,Mg_2SiO_4 + 3\,H_2O \rightarrow Mg_3Si_2O_5(OH)_4 + Mg(OH)_2 \tag{1}$$
$$\text{olivine} \qquad \text{added} \qquad\qquad \text{serpentine} \qquad \text{brucite}$$

$$2\,Mg_2SiO_4 + 2\,H_2O \rightarrow Mg_3Si_2O_5(OH)_4 + MgO \tag{2}$$
$$\text{olivine} \qquad \text{added} \qquad\qquad \text{serpentine} \qquad \text{removed}$$

$$3\,Mg_2SiO_4 + 4\,H_2O + SiO_2 \rightarrow 2\,Mg_3SiO_5(OH)_4 \tag{3}$$
$$\text{olivine} \qquad \text{added} \qquad\qquad \text{serpentine}$$

Equations (2) and (3) show that it is not possible to convert olivine to serpentine without addition of silica or subtraction of magnesia. If the excess magnesia is used to form brucite, then a dunite can be converted to serpentinite by addition of water only [Eq. (1)]. The average

MgO/SiO_2 ratio (in wt %) for dunites is ~ 1.23 and if serpentinization of a dunite is accomplished only by addition of water, this ratio should remain constant. For analyzed brucite-bearing massive serpentinites derived from ophiolite dunites, the MgO/SiO_2 ratio (in wt %) has been found to be ~ 1.23 (Coleman and Keith, 1971), but the sheared and monomineralic vein serpentinites have a MgO/SiO_2 ratio (in wt %) between 0.99 and 1.00. Thus it is possible that all three reactions [Eqs. (1), (2) and (3)] may be valid approximations of the chemical changes which may convert dunite to serpentinite. For ophiolite harzburgites containing both olivine and orthopyroxene, the average MgO/SiO_2 ratio (in wt %) is 1.02, much closer to that of monomineralic serpentinite. When the ratio of olivine to orthopyroxene is approximately 1:1, then serpentinite alone can be formed by simply hydrating a harzburgite:

$$Mg_2SiO_4 + MgSiO_2 + H_2O \rightarrow Mg_3Si_2O_5(OH)_4$$
$$\text{olivine} \qquad \text{orthopyroxene} \qquad \text{serpentinite}$$

Lower MgO/SiO_2 ratios are also characteristic of the lherzolites, and brucite is uncommon in their serpentinized equivalents. If the olivine/orthopyroxene ratio is greater than one, then brucite can be expected; alternately, silica may be added or magnesia removed to produce a monomineralic serpentinite.

The average amount of iron in dunites, harzburgites, and lherzolites is about 7% weight percent FeO, with olivine and orthopyroxene containing nearly equal amounts. Small amounts of iron are contained in chromite and clinopyroxene. The Fe^{3+}/Fe^{2+} ratios of serpentines reflect the changes in the activity of oxygen during their formation. The formation of awaruite (FeNi) in association with, or in the place of, magnetite in many serpentines demonstrates the extremely low activity of oxygen (Nickel, 1959; Page, 1967). During serpentinization the partitioning of iron among serpentine, brucite, magnetite and awaruite depends on the availability of O_2. The alteration of ophiolite peridotites to serpentinite requires a large amount of water. Serpentine minerals contain from 12 to 13.5% weight percent water. The amount of water required to serpentinize a peridotite completely is controlled by the original proportions of olivine, pyroxene, and plagioclase and by the mobilities of Mg or Si, or both.

New information concerning the source of fluids responsible for serpentinization has allowed for further speculation on the conditions leading to the alteration of ophiolite peridotite to serpentinite (Barnes and O'Neil, 1969; Wenner and Taylor, 1971, 1973, 1974). Stable isotope study of the D/H and O^{18}/O^{16} ratios in serpentinites from numerous localities show that δD values of of lizardite-chrysotile exhibit a latitudinal

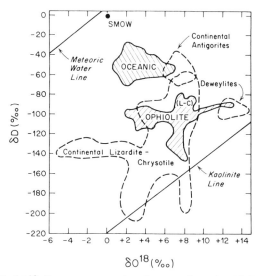

Fig. 40. δD–δO^{18} diagram comparing the oceanic and ophiolitic serpentines with other serpentines examined by Wenner and Taylor (1969). *L–C* lizardite-chrysotile. Note that the oceanic serpentines are clearly distinct from the continental and ophiolite serpentines. (After Wenner and Taylor, 1973)

variation that coincides with present-day variations recorded for mete-oritic waters (Wenner and Taylor, 1974). This variation has been interpreted to imply that some serpentinization producing lizardite and clinochrysotile is brought about by meteoric water in the earth's crust. Selected serpentinites from ophiolites exhibit this same pattern of δD variation with latitude and the δO^{18} has an extremely wide variation (Wenner and Taylor, 1973). However, a comparison of D/H and O^{18}/O^{16} ratios of ophiolitic serpentinites with oceanic serpentinites obtained by dredging showed that δD and δO^{18} of the oceanic serpen-tinites are very unique (Fig. 40). These different fractionation values indicate that ocean water modified by hydrothermal heating is probably the source of water responsible for the "oceanic" serpentinites whereas the great bulk of the serpentinized peridotites from ophiolites has been serpentinized by meteoritic waters at ambient temperatures up to about 300° C (Magaritz and Taylor, 1974; Wenner and Taylor, 1974). The implications of these stable isotope data require that most of the serpentinization of ophiolite peridotites must take place in a continental crustal environment rather than an oceanic environment. Assuming the ophiolite assemblage develops in the ocean, then it has to be tectoni-cally transported onto the continental margins prior to its complete serpentinization by meteoric water. Antigorite serpentinites have very restricted ranges of δD (-39 to -66) and δO^{18} ($+4.7$ to $+8.7$), suggesting that these serpentinites have formed during regional crustal meta-morphism in contact with waters developed as a result of metamorphism (Wenner and Taylor, 1973). The stable isotope ratios of the antigorites corroborate the earlier petrologic interpretation that antigorites formed at higher temperatures than chrysotile and lizardite.

The development of the large serpentinized peridotites is signifi-cantly linked with crustal processes that must be part of ophiolite tectonic evolution after the peridotite emplacement into orogenic zones. The change from a peridotite to serpentinite represents a great range in physical properties. Peridotites have a density of 3.3 whereas serpentinites average 2.55 and concomitantly the compressional wave velocity V_P decreases from ~ 8 km/s to ~ 5 km/s (Coleman, 1971b). Magnetite developed during serpentinization increases the magnetic susceptibility (K) and serpentinite masses are known to produce magnetic anomalies in excess of 500 gammas. Even though strength of unsheared serpentinites are comparable to massive igneous rocks (Raleigh and Paterson, 1965), estimation of the shear strength of completely tectonized lizardite-chrysotile serpentinites in New Idria, California by Cowan and Mansfield (1970) gives values of 1 bar. The weak nature of the tectonized serpentinite combined with its low density demonstrates that tectonic movement by plastic flow at low stress could easily account for the development of serpentinite mélanges. Thus, low temperature hydration of massive, relatively strong peridotites to sheared, weak serpentinites will have a drastic affect on the geometry of the original ophiolite assemblage in orogenic zones. The development of serpentinite within an ophiolite can be visualized as happening in three distinct situations: (1) Within the oceanic realm particularly along transform faults. (2) During tectonic transport into the continental margins. (3) As part of regional metamorphism. The source of the water can be fingerprinted by stable isotopes and demonstrates that serpentini-zation is a continuous process as long as meteoric, oceanic or meta-morphic water is available.

2.2 Rodingites

Metasomatism of varied rock types associated with serpentinites is widespread in Phanerozoic orogenic belts (Coleman, 1966, 1967; Dal Piaz, 1967, 1969). The metasomation is related to the process of serpentinization and tectonic history of the ultramafic emplacement (Table 12). Diverse rock types such as graywacke, gabbro, basalt, granite, dacite, and shale have been involved in the metasomatism. Originally, such metasomatized mafic rocks were called either *rodingites* or

Table 13. Typical analyzed rodingites from ophiolites

	1	2	3	4
SiO_2	40.2	43.2	41.4	36.52
Al_2O_3	10.5	8.6	21.6	15.12
Fe_2O_3	12.2[a]	11.8[a]	1.1	10.22
FeO	—	—	2.3	0.25
MgO	3.1	5.7	1.5	0.17
CaO	32.7	27.9	31.4	35.42
Na_2O	0.07	0.21	0.2	—
H_2O+	0.76	1.5	} 0.7	0.15
H_2O-	0.10	0.15		0.12
CO_2	0.01	0.03	—	—
TiO_2	0.56	0.63	0.1	1.92
MnO	0.15	0.22	0.2	—
Total	100.35	99.94	100.5	100.13
S.G.	3.59	3.44	—	—

[a] All Fe as Fe_2O_3.

(1) Eureka mine road, Oregon, metagabbro (Coleman, 1967). (2) Illinois River, Oregon, metagabbro (Coleman, 1967). (3) Balangero, near Torino, Italy, metasediment (Dal Piaz, 1969). (4) Testa Ciavara, near Lanzo, Italy, metagabbro (Dal Piaz, 1967).

garnetized gabbro, granatvesuvian fels, skarn, and ophiospherite. Through the years these metasomatic rocks have acquired the general term rodingite and for the purposes of this discussion I will use this as a general all-inclusive term.

Nearly all of the rodingite occurrences have similar characteristics. The metasomatic reaction zones are within or in contact with serpentinite and have not been observed in unserpentinized ultramafic rocks or in the high-temperature contact aureoles of such rocks. Rodingites are developed from diverse rock types associated with serpentinites. The reaction zones develop at the contact between tectonic inclusions, dikes and layered mafic rocks, and surrounding country rock and the serpentinite. The rodingites are characteristically involved in tectonic movements that have acted on the serpentinite. Synkinematic brecciation and mylonitization of some rodingites indicates that deformation may accompany metasomatism. The chemical changes recorded in the rodingites as a result of metasomatism all trend towards a similar bulk composition (Table 13). All rodingites are undersaturated with respect to silica and are enriched in calcium and to a lesser extent magnesium. Where the metasomatism has been complete many rodingites have the following approximate molecular proportion: $3\ CaO \cdot Al_2O_3 \cdot 2\ SiO_2 \cdot H_2O$. Rodingites derived from silicic rocks such as sandstone and granitic rocks, often contain a secondary feldspar-rich

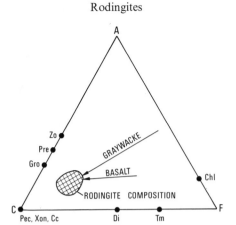

Fig. 41. Progressive changes in basalt and graywacke produced by calcium meta-somatism illustrated on an ACF diagram. Minerals characteristic of rodingites are also plotted: *Zo* Zoisite; *Pre* prehnite; *Gro* grossularite or hydrogarnet; *Pec* pectolite; *Xon* Xonotlite; *Cc* calcite or aragonite; *Di* diopside; *Tm* Tremo-lite-actinolite; *Chl* chlorite

core or zone away from the serpentinite contact. Chlorite and nephrite also may form within serpentinite at its contact with the rodingite. Rodingites are small-scale localized metasomatic occurrences and are not related to a regional metasomatism. Localized alteration zones which have rodingite chemistry have not as yet been reported from other geologic environments.

Hydrogarnet is the characteristic mineral of rodingites and is commonly associated with idocrase, diopside, prehnite, xonotlite, wollastonite, chlorite, sphene, and tremolite—actinolite (nephrite) where mafic igneous rocks have been metasomatized. In more silicic rocks modified by the metasomatism xonotlite, albite, potassium feldspar, and fibrous actinolite are commonly present.

The chemical changes brought about by rodingite metasomatism have a similar trend for all rock types, although the initial rock composition exerts a minor influence on the end product (Coleman, 1967) (Fig. 41). Rocks initially containing considerable amounts of silica and alkalis, may, after metasomatism, contain enriched areas of albite or potassium feldspar away from the reaction zone. Metasomatism in the reaction zone of both igneous and sedimentary rocks is characterized by enrichment in calcium and, in some rocks magnesium accompanied by an apparent movement of the alkalis away from the contact. Furthermore, the original unaltered rock always loses silica. The important point is that desilication and calcium metasomatism are universal in these altered rocks (Fig. 42).

Assuming that hydrogarnet is related to the breakdown of the original calcic plagioclase and does not result from the breakdown

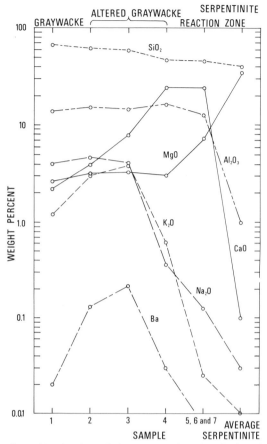

Fig. 42. Semilogarithmic plot of the progressive chemical variation of samples from Cape San Martin, California within a rodingitized graywacke. The abscissa does not represent horizontal distance between samples; points are placed equidistant to exhibit chemical variations. (After Coleman, 1967)

of ferromagnesians, which instead alter to chlorite and actinolite, a series of reactions can be written that may represent the approximate chemical nature of the metasomatism.

$$3\,(CaO \cdot Al_2O_3 \cdot 2\,SiO_2) + Ca^{+2} + 2\,H_2O \rightarrow$$
anorthite
$$4\,CaO \cdot 3\,Al_2O_3 \cdot 6\,SiO_2 \cdot H_2O + 2\,H^+.$$
zoisite

$$4\,CaO \cdot 3\,Al_2O_3 \cdot 6\,SiO_2 \cdot H_2O + 5\,Ca^+ + 13\,H_2O \rightarrow$$
zoisite
$$3\,(CaO \cdot Al_2O_3 \cdot 1.5\,SiO_2 \cdot 3\,H_2O) + 1.5\,SiO_2 + 10\,H^+.$$
hydrogarnet

Table 14. Chemical compositions of Ca^{+2}–OH^{-1}-type waters issuing from peridotites, concentrations in mg/l

		1	2	3	4	5
	pH	11.54	11.77	11.25	11.3	11.5
	T°C	20	18	31	28	25
	Ca^{+2}	40	53	35	120	60
	Mg^{+2}	0.3	0.3	0.1	0.2	0.1
	Na^{+1}	19	50	33	110	230
	K^{+1}	1.1	1.2	2.3	6	8
	Cl^{-1}	63	55	19	170	280
	SO_4^{-2}	0.4	0	0	4.8	8.6
	SiO_2	0.4	0.3	5.9	0.2	0.1
(Calc)	OH^{-1}	42.8	62.3	50.5	47.8	60.6
	Li^{+1}	—	0.02	0.02	0.02	0.01

Spring within: (1) Burro Mountain peridotite, Calif. (Barnes and O'Neil, 1969). (2) Cazadero peridotite, Calif. (Barnes and O'Neil, 1969). (3) Red Mountain peridotite, Calif. (Barnes and O'Neil, 1969). (4) Semail ophiolite, Oman (U.S.G.S., unpubl. data). (5) Semail ophiolite, Oman (U.S.G.S., unpubl. data).

These reactions illustrate the nature of the metasomatism as being driven to completion by an increase in the activity of calcium and water. Other reactions can be written to include diopside, prehnite, and idocrase, but in general the same relations hold. Calcium released from pyroxene during serpentinization is the main source of the metasomatic calcium. The average CaO content of 23 peridotites is 3.5 weight percent, and for 19 dunites, 0.75 weight percent (Poldervaart, 1955). The average CaO content determined from 26 analyzed serpentinites (Faust and Fahey, 1962) is 0.08 weight percent. These analytical data for CaO support the concept that calcium released during serpentinization becomes available for the metasomatic process.

Evidence for the presence of calcium release during serpentinization is also afforded by the occurrence of calcium hydroxide waters issuing from partly serpentinized peridotites (Barnes et al., 1967; Barnes and O'Neil, 1969; Barnes et al., 1972). These small sluggish springs have travertine aprons and discharge a calcium hydroxide-type (Ca^{2+}–OH^{-1}) water that has unusually high pH (11 +) values (Table 14). According to Barnes and O'Neil (1969, p. 1957): "The serpentinization process removes the CaO component and little else from the peridotites and dunites. In all Ca^{+2}–OH^{-1}-type waters are supersaturated with respect to diopside and tremolite. The Ca^{+2}–OH^{-1}-type waters are probably incompatible with (unsaturated with respect to) minerals in rocks surrounding the ultramafic rocks and may help explain the metasomatic rocks ('rodingites'). The metasomatic zones may be a product of serpentinization and diagnostic of the nature of the ultra-mafic rocks at the time of emplacement."

The tectonic nature of some serpentinite contacts and the asso-ciated low temperature reaction zones (rodingites) is strong evidence indicating that emplacement of the ultramafic parts of the ophiolite coincide with major tectonic events (Coleman, 1967; Dal Piaz, 1967). The widespread occurrence of rodingites within sheared serpentinites and serpentinite mélanges indicates that calcium metasomatism is a normal by-product of serpentinization. The pervasive presence of calcium hydroxide waters within serpentinized peridotites by analogy is very similar to the reaction that takes place in the formation of cement. Potentially, then wherever these calcium hydroxide waters encounter rocks higher in silica than the peridotites ($\sim 45\% \mathrm{SiO_2}$) reactions can take place whereby calcsilicate minerals will replace and invade the host whether it be an exogenous inclusion or endogenous rock within the peridotites. It is, therefore, important to recognize the "rodingites" as by-products of the serpentinization process and not the result of high-temperature contact phenomena related to an earlier igneous history of the ophiolite.

2.3 Hydrothermal Metamorphism

Many mafic rocks from ophiolite assemblages have undergone a "spilitic" metamorphism that appears to be widespread and uniform within the upper parts of the ophiolite (Gass and Smewing, 1973; Spooner and Fyfe, 1973). The mineral assemblages indicate a down-ward increasing thermal gradient from zeolite facies to greenschist facies and perhaps to low-grade amphibolite facies (Table 12). The hydrothermal metamorphism appears to be restricted to the pillow lavas, sheeted dikes, and upper parts of the gabbro, and the apparent metamorphic zonal boundaries are disposed subparallel to the original horizontal layers within the constructional parts of the ophiolite (Gass and Smewing, 1973). The metamorphosed mafic rocks retain their original igneous textures and exhibit only local shearing. The disposition of the hydrothermal metamorphism as well as its apparent shallow downward termination within the gabbros indicates a system controlled by circulating hot water. The lack of metamorphism in the gabbros relates to their impervious nature.

If the assumption is made that most ophiolites form at a spreading center under a cover of seawater and that high heat flow is maintained after their formation, it is possible to show that subsea geothermal systems operating at spreading centers are capable of producing this hydrothermal metamorphism (Spooner and Fyfe, 1973). High heat flow is symmetrically disposed on either side of present-day spreading centers (Lee and Uyeda, 1965) and within the ridge axis geothermal gradients from 500° to 1400° C/km have been postulated (Cann, 1970; Spooner and Fyfe, 1973). Even though considerable fluctuation in the heat flow is seen away from the ridge axis it has been suggested that thermal gradients of 150° C/km are maintained to at least 100 km away from the axis.

The implications of this situation have been quantified by Spooner and Fyfe (1973, p. 294): "Since geothermal systems probably occur along all the ridge systems it is clear that there may be massive interaction between seawater and oceanic crust. The scale of this process may be estimated by considering the heat transfer involved. Deffeyes (1970) has estimated that about 12.5 km^3/yr; of basaltic material appears at the ridge systems at the present time. The bulk of this, about 7–10 km^3 is considered to be intrusive under a shallow volcanic cover. The heat produced can drive subseafloor convection. Where 1 km^3 of basic magma ($\sim 3 \times 10^{15}$ g) cools from 1200° C down to 300° C heat is liberated corresponding to the latent heat of crystallization ($\sim 3 \times 10^{17}$ cals) and the heat capacity ($\sim 7 \times 10^{17}$ cals): in all about 10^{18} cals. To heat 1 km^3 of seawater to 300° C requires about 3×10^{17} cals. Hence 1 km^3 of cooled basic magma can heat 3 km^3 of seawater to 300° C. This phenomena is massive and must constitute the major process involved in cooling of the intrusive material of the oceanic crust."

It thus seems inevitable that new oceanic crust generated at a spreading ridge will have undergone hydrothermal metamorphism. The distribution and grade of metamorphism will be controlled by circulating seawater and the thermal gradient. Distribution of this ocean floor thermal metamorphism within on-land ophiolites from eastern Liguria, Italy (Spooner and Fyfe, 1973), Cyprus (Gass and Smewing, 1973) and Oman (Coleman, unpubl.) is similar in many respects. In all of these occurrences the zeolite facies assemblages are present in the upper parts of the sequence followed downward into greenschist facies where the effects of the hydrothermal alteration disappears in the upper parts of the layered gabbros (Fig. 44). Usually the metamorphic assemblage in the upper gabbros suggests low amphibolite facies and from this we can estimate a temperature of 500° C (Winkler, 1974). The vertical thickness affected by the hydrothermal metamorphism in Oman and Cyprus varies from 2 to 3.2 km respectively. Dividing this into a maximum temperature of 500° C gives an estimated thermal gradients from 150° C/km to 250° C/km. Such steep thermal gradients can be best explained by hot circulating sea water within the upper few kilometers of newly formed oceanic crust developed at a spreading center. Regional burial metamorphism

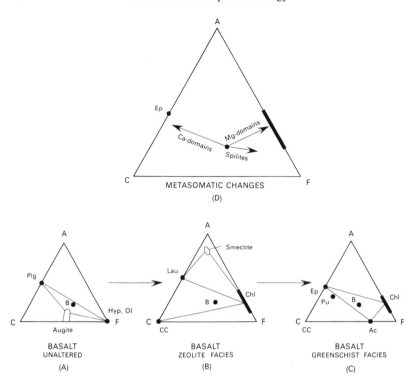

Fig. 43. Mineral assemblages in metabasalts undergoing hydrothermal seawater alteration from fresh basalt through zeolite to greenschist facies: $A \rightarrow B \rightarrow C$. Three possible trends of metasomatic alteration are shown in D. *Filled circle:* average basaltic composition. *Plg* plagioclase; *Hyp* hypersthene; *Ol* olivine; *Lau* laumontite; *Cc* carbonate; *Chl* chlorite; *Ep* epidote; *Pu* pumpellyite; *Ac* actinolite

thermal gradients are around 15° C/km and would require a depth of nearly 25 km to attain temperatures of 400° C. Therefore, ophiolites that had undergone regional burial metamorphism would perhaps be too thin to have such steep mineral isograds developed within their limited sequence.

Thermally metamorphosed basalts and diabase from ophiolites characteristically retain their primary igneous textures. Many of the arguments for primary spilite magmas are based on the fact that retention of textures is prima facie evidence for their igneous origin (Amstutz, 1974). However, careful studies of dredge hauls and drill cores from the oceans shows the presence of both "spilite" metamorphic assemblages as well as unaltered tholeiitic basalts (Melson and van Andel, 1966; Cann, 1969; Miyashiro et al., 1971; Miyashiro, 1972).

It is the opinion of the author that "spilite" metamorphism within the ophiolite sequence results from hot circulating H_2O within the upper parts of the newly-formed oceanic crust. Tectonic transport of these altered slices of oceanic crust onto the continental margins will in some instances juxtapose metamorphosed basalts and diabase against sedimentary country rock that has not been metamorphosed such as reported in Cyprus and Oman (Reinhardt, 1969; Gass and Smewing, 1973).

Within the zeolite zone of hydrothermal alteration, the igneous rocks are gray-brown to reddish-brown. The plagioclase is replaced by zeolites, with laumontite, analcime, and natrolite being the most common new minerals after the feldspar. Hematite and calcite are abundant, with calcite forming some veins and filling amygdules. Smectite is the common layered silicate in the zeolite zone and perhaps is the most abundant mineral within the glassy matrix of the altered basalts. Hyaloclastic material intersitial to the pillow lavas is also converted to smectite and iron oxide. In some ophiolite sequences the zeolite assemblages are not developed and chlorite, albite, and pumpellyite are common constituents. The downward progression from the upper zeolite assemblages into the greenschist assemblages is marked by gradational change (10–20 m) in color to light-green-blue (Gass and Smewing, 1973). Chlorite replaces smectite and albite replaces the igneous feldspar. Typically the igneous clino-pyroxene remains completely unaffected in the zeolite and lower grade greenschist facies alteration. Also as part of this transition, the calcite and hematite decrease in abundance and epidote along with magnetite make their appearance. Sphene becomes an abundant accessory and quartz appears in the groundmass imparting a hardness to the greenschist facies rocks. At deeper levels within the ophiolite, parti-cularly in the sheeted dikes, actinolite makes its appearance replacing clinopyroxene. Thus the typical greenschist assemblage of albite, epidote, actinolite, chlorite and sphene develops about 1 km below the pillow sediment interface. Within the gabbros, where the hydro-thermal metamorphism dies out, calcic plagioclase persists and coexists with actinolite and chlorite. At this level, the igneous clinopyroxene reacts being replaced by actinolite and chlorite. In these rocks albite and epidote are rare. It had been suggested earlier by Miyashiro et al. (1971) that calcic plagioclase and actinolite assemblages from the mid-Atlantic ridge may represent amphibolite facies metamorphism. This remarkably steep thermal gradient and the development of mineral zoning downward is nearly identical to the mineral zoning found in the altered mafic rocks within the Reykjavik geothermal field of Iceland (Sigvaldason, 1962). The significant difference between these two systems

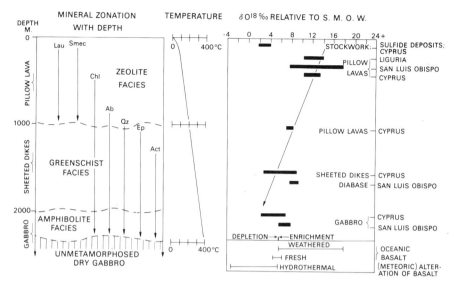

Fig. 44. Relationships between mineral zoning and depth in idealized ophiolite section that has undergone hydrothermal metamorphism. Assumed thermal gradient is 150° C/km. δO^{18} values taken from Spooner et al. (1974), Heaton and Sheppard (1974), Magaritz and Taylor (1976). *Lau* laumontite; *Smec* smectite; *Chl* chlorite; *Ab* albite; *Qz* quartz; *Ep* epidote; *Act* actinolite

is that the hydrothermal alteration of the ophiolite sequences is nearly uniform over large areas whereas the alteration of the geothermal field in Reykjavik is a local phenomena related to thermal springs.

The hydrothermal alteration of the upper parts of the ophiolite sequence requires the presence of circulating water. Evidence from on-land ophiolites indicates that the depth of this hydrothermal alteration is within the upper 2–3 km. Oxygen isotope geochemistry has provided important and indirect clues that the source of the water taking part in the hydrothermal alteration is sea water (Heaton and Sheppard, 1974; Spooner et al., 1974) (Fig. 44). Hydrothermal alteration of a rock by a fluid will change the oxygen isotope ratios of the rock away from the original igneous values. The average δO^{18} values of unaltered basalts is usually $+6\% \pm 0.5\%$ (Taylor, 1968). Continental examples of hydrothermally altered mafic igneous rocks show negative δO^{18} shifts and represent depletions brought about by interaction of the rock with hot water of meteoric origin (Taylor and Epstein, 1963; Taylor, 1971). In contrast to these depletions, the hydrothermally altered ophiolites from Mediterranean area are enriched in O^{18} relative to unaltered basalts by as much as 7% (Spooner et al., 1974). Even more extreme positive shifts are recorded for weathered basalts taken from deep ocean environments, where the fractionations took place at low temperatures. The δO^{18} enrichments of the upper parts of the ophiolite pass downwards into depletions within the higher grade metamorphic zones and is related to the increasing metamorphic temperatures at depth (Spooner et al., 1974). The stable isotope data indicates that the water circulating during the hydrothermal alteration of the ophiolites was sea water rather than meteoric water. The presence of seawater within the hydrothermal system responsible for the alteration of the ophiolites provides a potential fluid to bring about large scale seawater-basalt interactions. Evidence for such inter-action is found in the hydrothermally altered ophiolites where domains of contrasting composition may be present within a single exposure (Fig. 43d). There seems to be at least three important alteration domains: (1) Ca-enriched areas of monomineralic or bimineralic epidote and/or pumpellyite; (2) Spilite lithology where albite, chlorite, and calcium silicates dominate; (3) Mg-enriched areas that represent the complete alteration of glass to chlorite. These same alteration domains have been recognized in continental terrains where mafic volcanics have undergone burial metamorphism (Smith, 1968; Jolly and Smith, 1972; Vallance, 1974). Recent experimental work by Bischoff and Dickson (1975) on seawater-basalt at 200° C and 500 bars provides information on the possible species involved in this kind of hydrothermal alteration. They found that Mg was continuously abstracted from seawater and probably forms a Mg-rich chlorite or smectite from the glass. Ca precipitated as $CaSO_4$ early in the experiment but continued to increase in the seawater. Heavy metals, such as Fe, Mn, Ni, and Cu are leached under the conditions of the experiment and could concentrate in the fluid in high enough concentrations to be considered potential ore forming fluids. The results of this experiment point to at least a partial explanation of the alteration domains within ophiolites. The spilite assemblage must represent the residual part of the basalt that has undergone hot seawater reaction (Fig. 43d). Sodium enrichment may be related more to the loss of calcium and other elements rather than addition of sodium. Bischoff and Dickson (1975) found that potassium was strongly leached from the basalt during their experiment whereas sodium remained nearly constant in the seawater. The calcium-rich domains, such as the epidosites, must represent a major movement of calcium and alumina during the hydrothermal alteration. The low temperature formation of smectites (montmorillonites) and higher temperature formation of chlorite represents an important seawater action that removes Mg from the oceans. Hydrothermal brines developed during the alteration of the upper 3 km of oceanic crust (ophiolites) must then have the potential of forming metalliferous concentrations within the ophiolites or at the interface between ocean water and points of brine discharge such as are now taking place in the Red Sea (Bischoff, 1969). As was mentioned in earlier chapters, this hydrothermal alteration of the upper parts of the ophiolite sequence will have a profound affect on the primary chemistry of these igneous rocks. Thus, the comparison of various igneous suites within ophiolites with unaltered volcanics cannot provide definitive answers to the igneous history of these rocks.

3. External Metamorphism

3.1 Metamorphic Aureoles

Thin zones of high grade amphibolites are present at the base of several large obducted slabs of ophiolite (Allemann and Peters, 1972; Williams and Smyth, 1973; Glennie et al., 1974) and granulites are commonly associated with non-ophiolitic lherzolite tectonic massifs (Kornprobst, 1969; Loomis, 1972a, b, 1975; Nicolas and Jackson, 1972) (Table 12). The significance of these rocks has been given various interpretations over the past 50 years and up to the present time considerable controversy still surrounds their interpretation (Thayer and Brown, 1961; Thayer, 1967, 1971). Earlier opinions on the origin of the emplacement of peridotites focused on their intrusion as magmas and the associated aureoles were considered to be the result of a contact metamorphic effect on the country rock resulting from still hot ascending ultramafic magma (Smith, 1958; MacKenzie, 1960; MacGregor, 1964). Tectonic emplacement of peridotites was not, however, completely neglected (Irwin, 1964).

It is important to first separate the ophiolite sequences and their associated amphibolites from the non-ophiolitic lherzolite tectonite massifs that have associated granulites, as was recently done by Nicolas and Jackson (1972). The non-ophiolitic lherzolite massifs represent undifferentiated mantle from beneath the continents and the associated granulites are probably metamorphosed lower continental crust. Their emplacement is along deep fundamental faults initiated as mantle diapirs into the lower parts of the continental crust (Kornprobst, 1969; Loomis, 1972a, b, 1975). The associated granulites are usually more extensive than can be accounted for by contact metamorphism of a rising, solid, diapir of peridotite (Loomis, 1972b, p. 2492). The final tectonic emplacement of these peridotites into the crust distorts the original geometry of the peridotite and the granulites precluding a proper geologic evaluation of their formation. The non-ophiolitic lherzolite tectonites contain only peridotite in contrast to the ophiolite sequences where gabbros, diabase, and pillow lavas develop above the underlying peridotite. For the purposes of this discussion, the continental lherzolites of Nicolas and Jackson (1972) will be excluded.

The metamorphic aureoles associated with ophiolites are always located at the base of the peridotite and consist of narrow zones (usually less than 500 m) of amphibolite. The metamorphic fabric of the amphibolites show polyphase deformation with the second generation schistosities and fold axes parallel to the contact between the peridotite and amphibolite. Structures within the peridotite are also subparallel to the contact and suggest a recrystallization during the second generation of amphibolite deformation. Hornfels textures that are characteristic of static igneous contacts have not been reported as primary or secondary textures within these aureoles. In Newfoundland, downward progression from high grade amphibolites into greenschist assemblages at the base of the peridotite suggest extremely high thermal gradients (Malpas et al., 1973; Williams and Smyth, 1973). Within the Klamath Mountains of northern California a nearly continuous narrow belt (35 km × 0.5 km) of gneissic garnetiferous amphibolite occurs at the base of the Trinity ultramafic sheet (Irwin, 1964; Davis et al., 1965). The amphibolite, in the past, has been considered to be an intrusive contact aureole, but synkinematic deformation suggests that it may have developed during tectonic emplacement of the Trinity ophiolite. Metamorphic aureoles are not present at the base of all ophiolites, but often tectonic blocks of similar amphibolites are present in the mélange upon which the ophiolite is resting (Moores, 1969; Reinhardt, 1969). In California and Oregon serpentinite mélange units representing dismembered ophiolites often contain polyphase deformed amphibolites that are similar to those described as aureoles in Newfoundland (Coleman and Lanphere, 1971).

The narrow high grade amphibolite aureoles contain brown hornblende, clinopyroxene, garnet, calcic plagioclase in varying amounts. Usually within 10–15 m of the contact the mineral assemblage changes to essentially green-brown hornblende and calcic plagioclase and within 500 m the mineral assemblages are typical of the greenschist facies. Up to now there are no mineral analyses of coexisting phases and little can be said regarding the P–T conditions of formation. If we assume that the highest grade of metamorphism represented by the aureole rocks is between high-rank amphibolite and granulite then a temperature of between 600–700° C can be estimated. No realistic estimate of pressure can be made at this time.

The formation of such a narrow zone of metamorphic rock at the base of ophiolite slabs is a remarkable situation. The protolith for these narrow aureoles is difficult to establish since in most instances the tectonic transport of the ophiolite with its welded-on basal aureole is now resting on unmetamorphosed rock or a mélange. Williams and Smyth (1973) suggest that the aureole in Newfoundland was derived from fragmental mafic volcanic rocks formed along the continental margin and that a clear downward progression from amphibolite into only slightly metamorphosed mafic volcanics can be demonstrated. It must be noted here that the polyphase deformation of the high-rank aureole rocks from Newfoundland is not imprinted on the underlying mafic volcanic protolith and that there could exist low angle faults separating the high-rank and low-rank rocks under the aureole. In Oman, the high-rank amphibolite aureole of mafic composition apparently grades downward into metacherts and meta-shales; however, polyphase deformation of the Oman high-rank amphibolite is not present in the underlying lower rank metasediments, suggesting a cryptic discontinuity between the amphibolite and metasediments. The truly high-rank amphibolites from these aureoles appear, then, only as very narrow zones 10–20 m at the deformed base of the peridotite (Allemann and Peters, 1972).

It seems clear that these narrow aureoles are somehow related to the tectonic emplacement of ophiolites from the oceanic realm into or onto the continental margins or island arcs. Williams and Smyth (1973, p. 615) have provided, perhaps, the most viable mechanism for the development of these unique aureoles: "It is clear from several of the foregoing prerequisites that the west Newfoundland aureoles cannot represent typical hornfels of orthodox intrusive contacts. Particularly incongruous to such an interpretation are the special constant setting of the aureole, its narrow uniform width, its constant lithology, and its structural and metamorphic textural characteristics. However, a plate tectonic model that envisages the aureole as a result of the obduction of ocean crust and mantle onto a continental margin seems to fit most of the requirements. According to this model supracrustal rocks are overidden by subhorizontal sheet of forcefully expelled oceanic crust and mantle that is everywhere detached at approximately the same level, which in the case of the west Newfoundland examples is from 3 to 6 km within the mantle, that is, the thickness of the ultramafic unit of the ophiolite suite. The aureole would therefore evolve as a contact dynamo-thermal aureole and acquire its structural style and metamorphic zonation during the early stages of continuous, or possibly episodic, expulsion. Following the formation of the aureole, the ophiolite slice moved along a lower structural base so that the aureole was included as a structural underpinning welded against the basal peridotites. Subsequent transport was mainly effected by cold gravity sliding and the development of the characteristic mélange zones along which the sequences of transported slices in western Newfoundland were finally emplaced." Implicit in the model suggested by Williams and Smyth is a rather shallow detachment of the mantle where high heat flow allows thin slabs of oceanic crust to be emplaced. The high heat flow would provide the heat necessary to produce a thin aureole, perhaps, augmented by frictional heating produced during obduction (Church and Stevens, 1971; Church, 1972). The absence of amphibolite aureoles at the base of some ophiolites could be related to emplacement of cold slabs such as that in Papua New Guinea (Davies, 1971). Here the basal contact is against graywackes of the Owen Stanley Range which have recrystallized to high pressure blueschist assemblages.

Still another point of confusion arises when the gabbroic parts of the ophiolites have undergone metamorphism and take on the appearance of amphibolites related to the contact aureoles (Thayer and Brown, 1961; Thayer, 1967, 1972; Hutchinson and Dhonau, 1969; Hutchinson, 1975; Lanphere et al., 1975). Imbrication of tectonic slices of high-rank amphibolites representing lower crustal metamorphism with cold ophiolitic slabs of younger age along continental margins could also provide an alternate explanation (Coleman et al., 1976). Imbrication and tectonic translation of ophiolite slabs within the orogenic zones of continental margins virtually guarantees obliteration of the original geologic relationships developed during the formation of these aureoles. The following section will discuss these implications in light of metamorphism along continental margins.

3.2 Continental Margin Metamorphism

Nearly all oceanic lithosphere generated at spreading ridges eventually is carried to a consuming plate margin. Here it is destroyed by moving under the continental margin and finally sinking into the asthenosphere. However, within these consuming plate margins tectonic accidents transfer fragments of the oceanic lithosphere (ophiolites) into or onto the continental margins. The rheology of these consuming plate boundaries is now understood in a gross way and perhaps the most important feature is their asymmetry. This asymmetry produces a polarity in tectonic, metamorphic, and magmatic processes (Dickinson, 1971b; Grow, 1973; Ernst, 1974). In this section the discussion of external metamorphism of ophiolites will be related to the possible P–T regimes within a convergent plate boundary (Table 12). Metamorphic belts are characteristically formed along continental margins, island arcs or where orogenic zones developed during collision of two

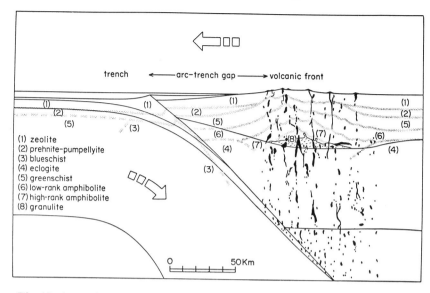

Fig. 45. Approximate spatial distribution of metamorphic facies in crustal rocks near an active convergent plate junction. The petrographic grid of Figure 46 has been employed with the thermal structure shown by Ernst (1974, Fig. 3). As in pressure-temperature space, the metamorphic facies boundaries are zones of finite appreciable volume within the earth. (After Ernst, 1974, Fig. 5)

continental blocks. Paired metamorphic belts characterize the continental and island arc metamorphic belts (Miyashiro, 1961; Ernst, 1974) but are lacking in the orogenic zones formed at continent-continent collision (Miyashiro, 1973b). The concept has been further refined by Ernst (1974) where the interplay between active subduction, calculated heat distribution, experimental petrology, and inferred geologic processes provide a three dimensional model of metamorphic facies at convergent plate boundaries (Fig. 45). From this model it is possible to consider the possible structural situation where oceanic lithosphere (ophiolites) could be metamorphosed. It also has to be assumed that decoupling or transfer of the oceanic lithosphere has to take place during or after the metamorphism in order for these rocks to be incorporated within the orogenic belts of the continental margins or island arcs.

It is possible to conceive of various P–T regimes that could be imprinted on oceanic lithosphere at consuming continental margins (Fig. 45). Within a cold sinking lithospheric slab the isotherms are deflected downward but pressure continues to increase during subduction and thus the ophiolites will undergo a progressive metamorphism from zeolite facies to the prehnite-pumpellyite facies and finally

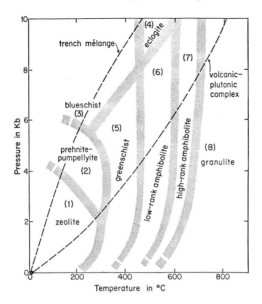

Fig. 46. Schematic metamorphic petrogenetic grid deduced from experimental—oxygen isotope data for common rock compositions. (After Ernst, 1973, Fig. 4; 1974, Fig. 4)

to the blueschist facies (Fig. 46). Thus a high-pressure, low-temperature blueschist metamorphism could be imprinted on the upper parts of an oceanic plate during subduction. There has not yet been found a completely preserved oceanic crustal slice that has undergone blueschist metamorphism with its stratigraphy intact. That is, there are numerous dismembered parts of ophiolite that have undergone blueschist metamorphism but they are presently exposed within a mélange or as highly deformed nappes.

The Zermatt-Sass ophiolites within the Piemonte zone of the western Alps are perhaps the largest exposed area of oceanic lithosphere that has undergone blueschist metamorphism as a result of subduction (Ernst, 1973; Dal Piaz, 1974). During Jurassic time the Piemonte basin formed and consisted, in part, of oceanic crust (Zermatt-Sass ophiolites). Subduction of this oceanic crust during the late Cretaceous (70–90 m.y.) converted the Zermatt-Sass ophiolites to a HP–LT assemblage consisting of blueschists and eclogites (Bearth, 1959, 1967, 1973). The Zermatt-Sass ophiolites are now part of a nappe consisting of tectonic slabs of ultramafics, gabbros, and pillow lavas showing a complete metamorphic and structural reworking (Bearth, 1974; Dal Piaz, 1974). The preservation of the Zermatt-Sass ophiolites from complete destruction in the asthenosphere was brought about by oblique planes of faulting within the subduction zone that produced a backward rise along the subduction plane to finally culminate in the emplacement of the Penninic and Austroalpine nappes (Dal Piaz et al., 1972). To further complicate the picture after emplacement of nappes, a greenschistic metamorphic event (Alpine) overprinted and irregularly destroyed the HP–LT assemblages within the Zermatt-Sass ophiolites. Ophiolites formed in the same basin within the Appennines were emplaced during the closing of the same oceanic basin by obduction and they only display the effects of oceanic metamorphism (Dal Piaz, 1974; Spooner et al., 1974). The complexities of the Alpine metamorphism of ophiolites in the Alps is further developed by Dietrich et al. (1974).

Blueschist metamorphism is characteristically developed in rocks underlying thick slabs of ophiolite that show oceanic metamorphism in their upper constructional parts such as in New Guinea and New Caldeonia (Coleman, 1972b). It has been suggested that tectonic overpressure developed during the emplacement of ophiolite slabs were instrumental in the development of the HP–LT blueschist metamorphism (Coleman, 1972b). Oxburgh and Turcotte (1974) have shown that blueschists may develop under thick overthrust sheets because of major perturbation of the regional thermal gradient—not tectonic overpressures. Where the overthrust sheets are in excess of 15 km thick, sufficiently high pressures and low temperatures could prevail long enough for blueschists to form and thus explain the presence of blueschists directly under thick ophiolite slabs. As shown in the previous section HT–LP, high-rank amphibolite can be more convincingly related to obduction of ophiolites than can blueschists.

The widespread occurrence of high grade blueschist and eclogite tectonic blocks of possible ophiolitic parentage within the Franciscan complex of California suggest that small parts of subducted ophiolite material escaped destruction and were transferred to the continental margins in mélange units (Coleman and Lanphere, 1971). Significantly there are no exposures of lithologically intact oceanic crust (ophiolite) within the Franciscan complex and those small metamorphosed fragments of oceanic crust that have undergone HP–LT metamorphism reveal tectonic transport within serpentinite mélanges (Coleman and Lanphere, 1971). There are, however, extensive areas of serpentinized ultramafics within the Franciscan complex that may have been involved in HP–LT subduction blueschist metamorphism. Since there are no pressure sensitive minerals within the serpentinized peridotites these dismembered parts of an ophiolite can be of no diagnostic value in reconstructing their metamorphic history. The actual volume of ophiolites that have undergone HP–LT metamorphism make up only a minor part of these metamorphic terrains giving credance to the concept that they represent tectonic accidents within steady-state subduction zones where most of the oceanic crust is destined for destruction within the asthenosphere.

It is of some interest to comment here that except for the Zermatt-Sass ophiolites and other minor Alpine localities most of the ophiolites emplaced in Alpine orogenic zones as a result of the closing of the Tethyan sea during the late Mesozoic have not been affected by HP–LT metamorphism. Furthermore, Pacific-type subduction during the late Mesozoic has incorporated only minor amounts of ophiolite in orogenic zones that have been affected by blueschist metamorphism whereas large tracts of sediments display HP–LT metamorphism. It can be concluded that most intact ophiolite sections are obducted onto the continental margin and escape high pressure metamorphism. Whereas oceanic crust and mantle is encorporated into the trench mélange only by tectonic accidents hence such subducted, then regurgitated ophiolites are both dismembered and exhibit high pressure metamorphism.

If we now turn our attention to the arc-trench gap and volcanic front areas at the continental margins as shown in Figure 45, it is possible to consider other pressure-temperature regimes which could affect ophiolites already incorporated into the continental edge. Stepping of the subduction zone seaward or the development of a volcanic-plutonic complex on the steeply dipping subduction zone could provide the heat necessary to pass into a prograde sequence from zeolite → prehnite-pumpellyite → greenschist → amphibolite. Thus it would be possible to convert an ophiolite to a HT–LP assemblage characteristic of the high temperature side of Miyashiro's (1961) paired metamorphic belts. The persistent association of amphibolites with metamorphic peridotites and lack of blueschists within the cores of progressive older orogens along Phanerozoic continental margins suggests that ophiolites of these older orogens may undergo HT–LP metamorphism.

Within the Klamath Mountains of northern California and southern Oregon, there are four different belts of ophiolite becoming sequentially younger oceanward (Irwin, 1973; Coleman and Irwin, 1974). In some instances the metamorphic peridotites from the Klamaths are closely associated with amphibolites that on close inspection reveal that they are metabasalts and metagabbros derived from the ophiolite. High angle faults juxtaposed the metaophiolites against island arc volcanics and sediments that have undergone zeolite, greenschist facies, or rarely blueschist metamorphism. The high-rank amphibolite facies metamorphism of these ophiolites indicates that they may have been involved in an island arc volcanic-plutonic complex where they form the basement rocks. Continued imbrication and compression along the

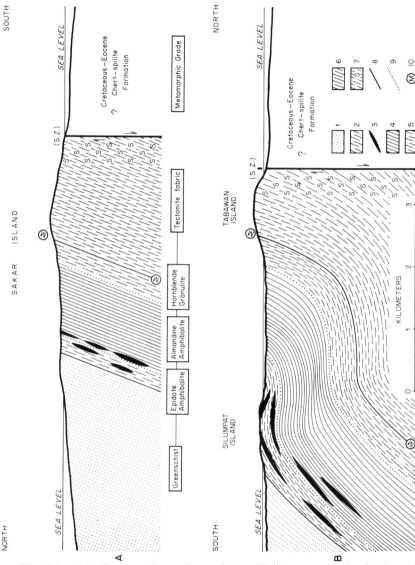

Fig. 47A and B. Cross sections of complete ophiolite sequences in the Darvel Bay-Labuk-Palawan ophiolite belt of North Borneo and Philippines. Sections are based on studies of outcrops on Darvel Bay Island of Tabawan, Silumpat, and Sakar. Upper layer of greenschist facies amphibolite is overlain by spilitic lava. Patterns: *1* greenschist facies fine-grained amphibolite (metabasalt); *2* epidote amphibolite-banded metagabbro; *3* amphibolitized dolerite sills; *4* almandine amphibolite—facies banded hornblende—plagioclase gneiss (metagabbro); *5* hornblende granu-lite-facies pyroxene—hornblende—plagioclase gneiss (metagabbro); *6* harzburgite, peridotite, and dunite with tectonite fabric; *7* serpentinized ultramafite; *8* sharp, conformable contact; *9* gradational metamorphic boundary; *10* Mohorovicic dis-continuity; *S.Z.* shear zone. (After Hutchinson, 1975)

continental margin produced high angle reverse faults that brought slices of the basement metaophiolites in contact with the lower metamorphic grade island arc volcanics and sediments.

Hutchison (1975) describes complete ophiolite sequences from the Darvel Bay-Lubuk-Palawan belt of North Borneo and Philippines that show HT-LP metamorphism from greenschist through high rank amphibolite facies into a metamorphic peridotite (Fig. 47). This sequence shows strong dynamothermal metamorphic fabric and even though Hutchison (1975) considers this as ocean floor metamorphism, it seems more likely to represent a metamorphic belt formed under an old island arc within the Sulu sea during Mesozoic times and later exposed by obduction of the metamorphosed arc basement onto North Borneo during late Cretaceous or early Tertiary.

Similar amphibolites derived from metamorphosed basalts and gabbros (ophiolites) have been reported from Yap (Shiraki, 1971), the Solomon Islands (Coleman, 1970), Puerto Rico (Tobisch, 1968), of Yugoslavia (Pamic et al., 1973). The recurrence of these amphibolites with metamorphic peridotites further illustrates the total metamorphism of ophiolites within the high heat flow regime of continental edges or island arcs. The important aspect of this type of metamorphism is that as long as the peridotites remain relatively dry, their deformation or recrystallization cannot be distinguished from that produced by HP–LT metamorphism in a dry environment. Confusion then arises when blueschists are associated with the same kind of peridotites as the amphibolite. Tectonic separation after upper level serpentinization of peridotite from its metamorphosed mafic assemblage produces a confusing mixing of amphibolites with blueschists as part of serpentinite mélanges (Coleman and Lanphere, 1973),

To confuse the metamorphic history of ophiolites further it is not uncommon to have younger granitic plutons invade tectonically displaced metamorphosed ophiolites and overprint them with a steep contact metamorphic thermal gradient (Evans and Trommsdorf, 1970; Dungan and Vance, 1972; Trommsdorf and Evans, 1972). As mentioned earlier the presence of amphibolites along the basal contacts of ophiolites has been interpreted as a dynamothermal aureole related to the emplacement of the ophiolite (Williams and Smyth, 1973). However, it is entirely possible that tectonic movement of an ophiolite peridotite over a regionally developed terrain of amphibolite could detach slices of amphibolite. These detached tectonic amphibolites could then conceivably come to rest at the base of the peridotite, and manifest all the aspects of an aureole. Considerable data therefore are required before amphibolites associated with allochthonous ophiolites can be properly interpreted.

Part V. Ore Deposits in Ophiolites

1. Introduction

Perhaps one of the most significant aspects of the new theories regarding the origin of ophiolites is the resurgence of interest (1972–1975) by mining companies in exploration for stratabound ore bodies within superficially worked out prehistoric and historic mines in copper-rich massive sulfide zones because of recognition of the geologic setting and localization of these deposits within ophiolite belts (Sillitoe, 1972; Holmes, et al., 1975; Huston, 1975). The fact that these massive sulfide deposits are stratabound within the pillow lava sections of the ophiolite sequence indicates that they probably formed within the ocean basins and that they are related to volcanogenic processes at spreading centers (Sillitoe, 1972, 1973; Duke and Hutchinson, 1974; Spooner et al., 1974). Differentiation of mafic and ultramafic magmas deep within the spreading centers also concentrate chromite and minor sulfides and these processes are equally important to the formation of ore deposits within the oceanic crust. There are, of course, other massive sulfide deposits formed in geologic environments quite distinct from those deposits occurring in ophiolites. The discussion to follow is concerned only with those massive sulfide deposits from the volcanic parts of ophiolites.

These ore deposits that formed within the oceanic realm are later emplaced as oceanic crust fragments within the orogenic belts of continental margins. Reconstructed oceanic crust stratigraphy shows volcanics at the top with associated metalliferous sediments which grade downward into sheeted dikes and then into a cumulate section whose base is usually a metamorphic peridotite. This stratigraphy provides a framework for systematic prospecting of on-land ophiolites. The massive sulfide deposits seem to be restricted to the extrusive volcanics, minor sulfide concentration are found in the cumulate gabbros, and chromite deposits are confined to the ultramafic parts of the ophiolite. Prior to the ideas generated by plate tectonics, mining geologists could not develop a unified geologic model in which to place their detailed observations from individual mining districts situated within submarine volcanics. However, in recent years, systematic exploration within the submarine volcanic sections of ophiolites has produced significant new copper-rich deposits (Holmes et al., 1975; Huston, 1975).

Transfer and incorporation of large slabs of oceanic crust onto the continental margin also produces, as a secondary product, another class of ore deposits that are equally as important as the magmatic and volcanogenic ore deposits. Where exposed areas of ophiolite peridotite have undergone tropical weathering, nickel and iron laterites are common, and under certain conditions of serpentinization asbestos deposits may form within the peridotites. The discussion that follows will provide a general background on the various ore deposits characteristic of the ophiolite assemblage.

2. Massive Sulfides

Massive sulfide bodies are situated within the pillow lava sections of many ophiolites. These sulfide bodies have certain common features, such as, a tendency to be stratabound and to occupy stratigraphic horizons within the volcanic section. The surrounding volcanic rocks have typically been affected by ocean floor hydrothermal metamorphism and the massive sulfides are found in those rocks that have undergone zeolite or greenschist grade metamorphism under steep thermal gradients (Gass and Smewing, 1973). Nearly all of these deposits have well-developed gossans consisting of bright colored iron oxides, hydroxides, and sulfates which attracted the ancient miners. Remnants of ancient slag heaps are present in many of these massive sulfide mining districts particularly in Cyprus and Oman (Bear, 1963a; Huston, 1975) and radio-carbon dating of charcol indicates that some of this ancient mining activity extends back to 2500 B.C.

In Cyprus, the massive sulfides are contained in the upper parts of the lower pillow lava series (see section on Troodos ophiolite, Cyprus) which consists primarily of pillow lavas and less common diabase dikes. Where the relations are preserved and not obscured by surface gossans, the massive sulfides are overlain by the Ochre Group, an iron-rich sediment containing sulfides but very low in manganese (Searle, 1972; Constantinou and Govett, 1973). These iron-rich sediments are very similar to the metal-rich muds that occur in the Red Sea axial deeps that have formed along the present-day spreading axis of the Red Sea (Bischoff, 1969; Amann et al., 1973; Bäcker, 1973). Unconformably overlying the lower pillow series is the upper pillow lava series considered to be an off-axis volcanic series and barren of any massive sulfide deposits (Gass and Smewing, 1973). The upper pillow lava series is uncomfortably overlain by the Perapedhi Formation of early Cretaceous age. The basal unit of this formation consists of ferromagnesian sediments (umbers) showing high concentrations of iron, manganese, arsenic, and copper and forming intermittant lenses as much as 100 m long and 5 m thick in depressions within the top surface of the upper pillow lavas (Elderfield et al., 1972; Robertson and Hudson, 1973; Robertson, 1975). These sediments are similar to those described by Anderson and Halunen (1974) and McMurtry and Burnett (1975) from the Bauer deep and have no apparent relationship with the massive sulfide deposits at the top of the lower pillows. The geologic field evidence on the Cyprus massive sulfide deposits demonstrate that these deposits developed after the extrusion of the lower pillow lavas and at the lava-sea water interface before the upper pillow lava-axis sequence formed.

In Oman, the massive sulfide deposits are concentrated in the upper parts of the pillow lava sequence (Bailey and Coleman, 1975). However, in contrast to those described in Cyprus, the massive sulfides are distributed irregularly below the upper parts of the pillow lavas. Thin lenses of iron-rich sediment are interbedded within the upper parts of the pillow sections and are similar to the ochres of Cyprus. However, these iron-rich sediments apparently do not form a cap above the massive sulfides as they do in Cyprus. The associated volcanics within the pillow lava section of Oman show a regional alteration to zeolite and greenschist grade metamorphism; however, there is no indication that this alteration was more pronounced around the massive sulfide occurrence. Sometime shortly after the development of the Oman ophiolite (~ 85 m.y.) it was deformed and parts of it underwent a lateritic weathering prior to the invasion of a shallow water Maestrichian carbonate sequence (Glennie et al., 1974). Distribution of the massive sulfides within the upper parts of the pillow sequence rather than at the very top as described from Cyprus indicates that the massive sulfides as well as the iron-rich sediments were not restricted to a specific stratigraphic level.

The massive sulfide deposit (York Harbour) from the Bay of Islands ophiolite occurs near and within the contact between a lower group of volcanics that have undergone extensive alteration in comparison to the overlying less altered lavas (Duke and Hutchinson, 1974). The stratigraphic control of the sulfide mineralization is very strong at this locality as all important occurrences are within the lower volcanic unit. Greenschist grade metamorphism has affected both the upper and lower volcanic units and is probably a result of ocean floor hydrothermal metamorphism (Williams and Malpas, 1972). Iron-rich sediments (ochres) are not reported from this deposit but red cherts within pillow interstices of the upper volcanic unit indicate some pelagic sedimentation associated with volcanism. Upadhyay and Strong (1973) have found that the massive sulfides within the Betts Cove area of Newfoundland near the Bay of Islands complex are concentrated at the top of the sheeted dike swarm within the pillow lavas. Here the massive sulfides have formed in the early stages of submarine volcanism rather than as in Cyprus where the massive sulfides apparently formed in the final stages of volcanism.

Ophiolites in the Italian Apennines contain massive sulfides in the pillow lavas just above the massive diabase sections as sheets or lenses (Holmes et al., 1975). And in this area there is reported extensive ocean floor hydrothermal metamorphism within the pillow lavas and diabase (Spooner and Fyfe, 1973). Metal-rich sediments (umbers) are commonly developed in the chert sediments that directly overlie the basalts of the Apennine ophiolite sections. These umbers are nearly identical to those described from Cyprus and indicate processes related to hydrothermal activity at spreading centers (Bonatti, 1975).

Similar massive sulfide deposits have been reported from the Paleozoic ophiolites of southwestern Japan, the Cretaceous-Paleogene pillow lavas in the Philippines, the Paleozoic greenstone belts of the Urals, U.S.S.R., and the ophiolites within Turkey (Sillitoe, 1972). It is of interest to note here that massive sulfides have not yet been reported from many of Paleozoic to Mesozoic ophiolite belts within the circum-Pacific margins. This may result from lack of exploration plus the fact that gossans may not be well developed in these regions.

All of the Phanerozoic ophiolite massive sulfide deposits consist predominantly of pyrite (>95%) with significant amounts of chalcopyrite, sphalerite, marcasite, and only minor amounts of pyrrhotite, galena, Au and Ag. Quartz, gypsum, chlorite and various kinds of sulfates are the common gangue minerals. The main parts of the massive sulfide bodies often show sedimentary banding and colloform structures suggestive of primary sedimentary deposition. These structures are characteristically disturbed by brecciation and slumping, with chalcopyrite infilling fractures and voids within these broken massive sulfides. Some chalcopyrite is present as inclusions within the pyrite but most of it is later than pyrite. A stockwork zone commonly underlies the massive sulfide bodies where brecciation of the volcanics is common and is cemented by quartz and pyrite. Here the volcanics are strongly altered (propylitization) with all ferromagnesian minerals replaced by pyrite suggesting that these stockwork zones represents channelways for circulating ore-forming brines (Constantinou and Govett, 1972, 1973; Searle, 1972). A generalized diagram of the massive sulfide deposits from Cyprus is given in Figure 48 and shows the various zones recognized

Genesis	Mode of formation	Mode of deposition	Ore type	Geol. col-umn	Ore zones	Mineralogical description	Type of mineral-ization	Minerals other than pyrite	Average Ore grade %S	%Cu
Magmatic	Oxidation / Sedimentary	Exhalative-sedimentary	Sulphides and iron mudstone	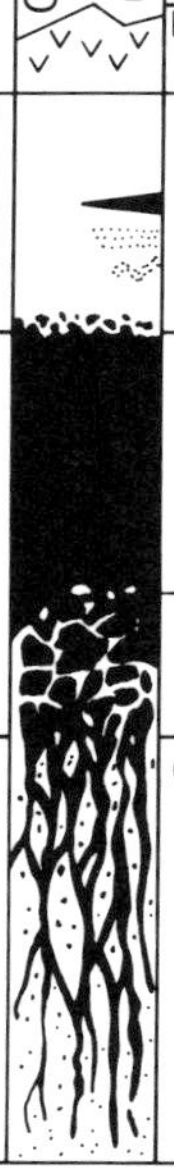	Post-mineral-ization lava Zone 1	Iron-rich and siliceous mudstones; oxidation of sulphides; lenses of friable black ore. Graded bedding, with grains of pyrite, and slumping boulders of colloform sulphide	Type 1	Sediment: goethite, silica, montmorillonite. Sulphides: covellite, sphalerite, chalcopyrite, marcasite	45–50	0.5–1.5
	Fumarolic		Massive sulphide		Zone 2	Fragmentary ore consisting of angular and hard blocks of yellow ore in matrix of black sandy ore	Type 2	Marcasite, sphalerite, po(v.r.), bn(v.r.), galena (v.r.)*, chalcopyrite	40–45	1.0–4.0
					Zone 2A	Fragmentary sulphide, blocks of pyrite in siliceous matrix with decreasing amount of black, friable sulphide	Type 3	Marcasite (r), chalcopyrite, sphalerite (r), po (v.r.) quartz, jasper, chalcedony	30–40	0.5–1.5
		Hydrothermal	Stockwork, cavity, fracture-filling, disseminations		Original lava surface Zone 3	More coarsely crystalline splendent pyrite as cavity-filling with much quartz and jasper	Type 4	Chalcopyrite, sphalerite, quartz, illite, jasper, chlorite	15–30	0.4–1.2
						Discrete fracture fillings with pyrite and quartz and jasper. Disseminations of pyrite in vesicles and cracks	Type 5	Chalcopyrite (r) sphalerite, (v.r.), rutile (r.), illite, quartz, chlorite, jasper	5–15	0.2–0.5

* From Agrokipia B only   r, rare   vr, very rare

Fig. 48. Summary descriptive data of Cyprus sulfide deposits. (After Searle, 1972)

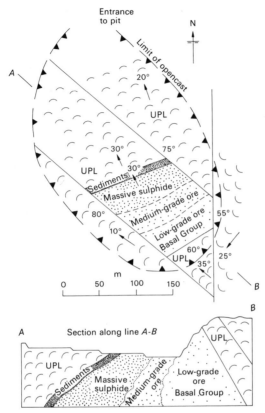

Fig. 49. Geologic sketch map and section of North Mathiati mine, Troodos ophiolite, Cyprus. (After Searle, 1972)

from the past mining operations. Documentation of this kind is not yet
available from other massive sulfide districts but the general descriptions
of other deposits now available indicate great similarity.

There is no compelling evidence of large scale replacement within
the ophiolite massive sulfide bodies and there is a general consensus
that these stratabound sulfides represent hydrothermal deposits of
metalliferous sulfide-rich sediments formed at the basalt-saltwater
interface (Sillitoe, 1972, 1973; Constantinou and Govett, 1973; Bonatti,
1975). The apparent fault control clearly shown by the Cyprus and
Oman deposits may represent localization of hydrothermal brines
along favorable zones of movement. Most of these faults do not penetrate
upwards into the overlying unconformable marine sediments and are
probably formed within the oceanic realm after the formation of the
volcanics at a spreading center (Gass and Smewing, 1973), (Fig. 49).

Spooner and Fyfe (1973) suggest the following sequence in the formation of these massive sulfide deposits: "... the stockwork may be interpreted as the product of prolonged precipitation from hydrothermal solutions which moved upwards in the discharge zone of a sub-seafloor geothermal system, after intense high temperature leaching at depth. The stratiform body may be identified as a heavy metal enriched sediment produced by inorganic precipitation under low oxygen fugacity conditions. Since each deposit may be identified as the discharge zone of a specific system, the model explains their sporadic occurrence." Evidence of pervasive circulation of hot seawater as shown by stable isotopes studies (see section on hydrothermal ocean floor metamorphism) in the upper parts of the ophiolite sequence near a spreading ridge provides an energy source and fluid capable of forming the massive sulfide deposits (Corliss, 1971). This concept is still somewhat controversial but has provided the impetus for exploration programs for massive sulfide deposits within ophiolite volcanics and the success of these programs provides evidence of the concept's validity.

3. Chromite

As was discussed earlier (sections on metamorphic peridotites and cumulate complexes) chromite (spinel) is a common accessory mineral in all of the various rock types found within the metamorphic peridotites and cumulate peridotites. In some instances the chromite becomes concentrated as tabular, pencil shaped, or irregular masses to form economic concentrations of chromite. The general term "podiform chromite" has been used to describe this class of deposits within peridotites of the ophiolite assemblage (Thayer, 1964). The occurrence of chromite deposits in large non-ophiolitic layered ultramafic and mafic intrusions such as the Stillwater complex in Montana and the Bushveld Complex in South Africa have been referred to as "stratiform chromite" (Jackson, 1961; Thayer, 1964). This distinction is a useful one and will be adopted in this discussion; it will be assumed that all of the chromite deposits in ophiolites are of the podiform type. As pointed out by Thayer (1964, 1969a, 1970), the podiform chromite deposits almost always have undergone some kind of metamorphic deformation and should be considered as metamorphic rocks whereas the stratiform deposits exhibit primary cumulate structures and textures characteristic of sedimentary rocks.

Most occurrence of podiform chromite deposits are associated with dunite rather than harzburgite or pyroxenites. There seem to be two separate modes of occurrence for the chromite deposits. Most commonly there is a nonsystematic distribution of the podiform deposits within metamorphic peridotites with no relationship between size of the deposit and its enclosing dunite. These deposits have a strongly super-imposed metamorphic fabric and the shape of the body may be highly contorted by folding or compressed into tabular or pencil-form deposits (Fig. 50).

Another class of chromite deposits are those that have a very close spatial relationship between the layered gabbro and peridotite contact within the ophiolite sequence. The shape and form of the chromite bodies related to the gabbro are similar to those that are found in random distribution within the metamorphic peridotites. The large chromite deposits in northern Luzon, Philippines, the Camagüey deposit of Central Cuba, and those from northern Oman show a close spatial relationship with gabbro contact (Flint et al., 1948; Thayer, 1964; Greenwood and Loney, 1968). The position of the chromite ores within 100 or 200 m of the layered gabbros in Oman and well within the recognizable cumulate ultramafic parts of the layered gabbros indicates that these chromites probably developed as magmatic segregations in the early stages of crystallization.

There is no clear consensus regarding the significance of the distribution of podiform chromite within the ophiolite peridotite. Because of the deep mantle deformation characteristics of the peridotites it seems entirely plausible that most of these chromite bodies have undergone intense dislocation along with folding and general disruption of the primary structures.

Textures of the podiform chromite ores are quite variable ranging from massive interlocking grains to euhedral crystals forming distinct layers. In many cases the individual grains show a pull-apart lineation related to their metamorphic deformation (Thayer, 1964). Gradded bedding similar to the stratiform chromite ores is often encountered but invariably shows folding and deformation (Thayer, 1970). The basic pattern of the textures in the podiform deposits appears to be a disruption of primary cumulate chromite structure by deep seated mantle subsolidus deformation. The large size of the chromite crystals and the scale of the chromite segregations that are randomly distributed within the metamorphic peridotites certainly suggests that the first stage of chromite formation was related to a deep mantle magmatic segregation of chromite, an event unrelated to the formation of the presently overlying oceanic crust.

The composition of the chromites from various podiform deposits is highly variable as is the accessory chromite from the peridotites (Irvine, 1967; Dickey, 1975) (Fig. 51). This high degree of variability reflects in part the complicated mantle history these deposits have

A   KANDAK-
      KARABÖRTLEN (IX)

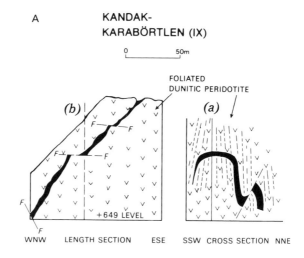

WNW   LENGTH SECTION   ESE   SSW CROSS SECTION NNE

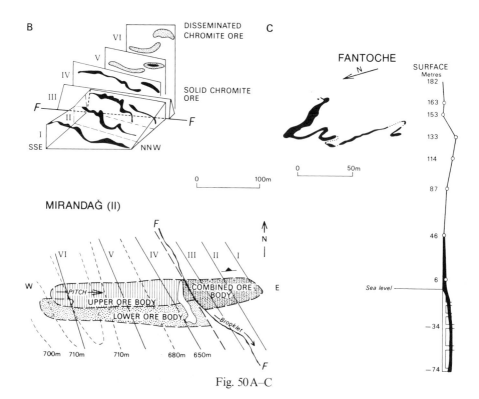

Fig. 50A–C

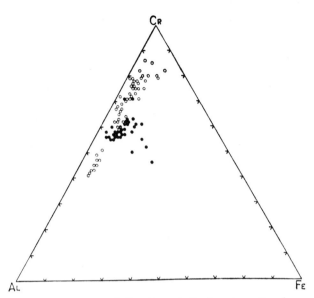

Fig. 51. Atomic proportions of Cr, Al, and Fe for chromites from stratiform intrusions *(solid circles)* and podiform deposits *(open circles)*. Fe^{3+} calculated from total Fe by assuming spinel stoichiometry and shown as Fe on diagram. Analyses by electron probe. (After Dickey, 1975)

undergone. Stratiform chromite compositions can be tied directly to their position in the differentiation sequences; Fe increases with the degree of differentiation whereas Cr_2O_3 content decreases (Jackson, 1963). The iron content of podiform chromites remains fairly constant (16% Fe) but Cr_2O_3 and Al_2O_3 show a reciprocal relationship and most plots of podiform chromites give a bimodal distribution: high chromium deposits and high aluminium deposits (Thayer, 1970). The reasons for this bimodality is not clear but experimental data show that high alumina chromites are more likely to develop under higher pressures than high chromium chromites assuming a constancy in the bulk composition of the deposits (Irvine, 1967) and as not earlier (section on metamorphic peridotites), spinels from the high temperature lherzolites seem to be chromium poor and alumina rich. In Oman, the chromite deposits situated within 100–200 m of the

Fig. 50A–C. Structures and shapes of podiform chromite deposits from ophiolites. (A) Vertical sections of Handak-Karaboren chromite deposit from Turkey. (After Thayer, 1964, and modified by van der Kaaden, 1970). (B) Vertical sections and plan view of the Miranday chromite deposit near Harmancik, Turkey. (After van der Kaaden, 1970, modified from Henckmann, 1942). (C) Horizontal and vertical plan of Fantoche chromite mine, situated within the Tiebaghi massif, New Caledonia. (After Routhier, 1963, p. 208) gabbro contact are high chromium chromites whereas the podiform deposits within the lower parts of the metamorphic peridotite are high alumina chromites. Nearly all of the high alumina and high chromium chromite deposits come from the podiform deposits within ophiolite complexes. The complex mantle history recognized in the depleted, metamorphic harzburgite and dunites, the most common host rocks for podiform chromites, demonstrates that these rocks have probably undergone a number of deformations within the mantle prior to their emplacement as part of an allochthonous slab. Crustal tectonism, metamorphism, and serpentinization all effect the original nature of the enclosed chromite deposits. It is clear however that the podiform chromites are formed as part of an early magmatic history within the upper parts of the mantle and that their present shape and form is a result of a series of mantle and crustal events (Thayer, 1969a). All of these geologic events conspire to produce an incredibly complicated setting for chromite ore deposits which leads to confusion in establishing guides for chromite prospecting and mining.

The most important chromite deposits of the world, exclusive of the Precambrian stratiform deposits, are contained in allochthonous ophiolite bodies within the Phanerozoic orogenic fold belts (Engin and Hirst, 1970). In the Alpine fold belt of Europe and the Mideast, important chromite deposits are found in Yugoslavia and Greece (Hiessleitner, 1951), Turkey (Engin and Hirst, 1970; van der Kaaden, 1970), Iran (Haeri, 1961), and Pakistan (Bilgrami, 1964). In the circum-Pacific area, the big chromite deposits are directly tied to the distribution of the ophiolite belts and are found in the Philippines (Rossman et al., 1959), New Caledonia (Caillere et al., 1956), Oregon (Thayer, 1940; Ramp, 1961), Borneo (Hutchison, 1972, 1975). In the Caribbean area, the ophiolite belts of Cuba contain the most important deposits (Thayer, 1942, Flint et al., 1948).

4. Deposits Formed by Secondary Processes

4.1 Laterites

The large allochthonous slabs of ophiolite peridotite emplaced along continental margins in many cases exhibit an upper surface of subdued topography with extensive flat surfaces (erosional surfaces?) Laterites form on these surfaces where they have been exposed long enough to a tropical or high rainfall climate during tectonically stable periods (Fig. 52). In this particular situation, leaching of the peridotite is especially effective and leaves behind certain insoluble

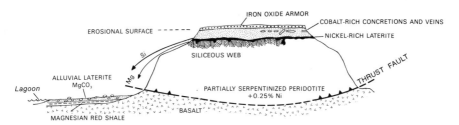

Fig. 52. Geochemical scheme of lateritic alteration of peridotites in New Caledonia showing relationships between ferruginous laterite and nickel concentration. Observations by J. Avias and P. Routhier. (Modified after Routhier, 1963, p. 208)

compounds. Iron and nickel become concentrated in the process of lateritization and are found in the leached red to yellow soils that develop on these surfaces. Vast tonnages of these laterites are distributed on the weathered surfaces of ophiolites in the tropical regions of the earth and represent the largest available reserves of low-grade nickel deposits. Although the iron-laterites also make up a huge reserve of low-grade iron, their importance is secondary as they cannot compete with the higher grade-ore of the large Precambrian sedimentary iron-ore deposits.

Nickel and iron laterite deposits are found in Cuba (Kemp, 1916; De Vletter, 1955), Guatemala, Colombia, Oregon (Hotz, 1964), New Caledonia (Chetelat, 1947; Routheir, 1952), Indonesia (Reynolds et al., 1973), Oman, Philippine Islands, New Hebrides (Coleman, 1970) and the Solomon Islands (Coleman, 1970). The laterites are almost invariably developed on partially serpentinized peridotites or on unsheared completely serpentinized peridotites. Laterites have not been reported on highly sheared and tectonized serpentinites because serpentine is much more stable than olivine in the weathering environment. The laterites can be placed into two broad types (Hotz, 1964). (1) Those deposits developed on serpentinized peridotite such as those in Cuba and the Philippines are called nickeliferous ferruginous laterites and contain approximately 40% Fe and average about 1% Ni. These deposits represent large reserves of low grade iron-ore and the nickel is highly dispersed. (2) The second type of deposit is typically developed on peridotites that are weakly serpentinized such as those from New Caledonia and Oregon and are referred to as nickel silicate deposits (Hotz, 1964). The nickel silicate deposits are relatively low in iron (<35% Fe) and contain garnierite (nickel silicate ~15% NiO.) as the main ore mineral in the lower part of the weathered zone. The nickel silicate deposits of New Caledonia average 3.5% Ni and certain zones rich in garnierite may run as

Weight Percent

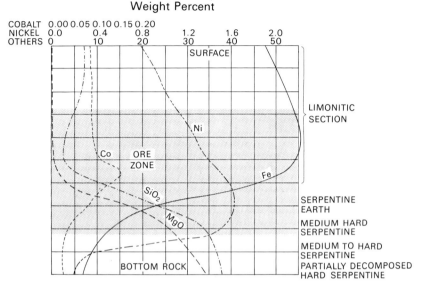

Fig. 53. Graphic representation of the distribution of certain selected elements in nickel laterite ore from Ocujal ore zone, Cuba. (Modified after de Vletter, 1955)

high as 10% Ni (Chetelat, 1947) whereas those from Oregon average about 1.5% Ni (Hotz, 1964). These two types of laterite deposits usually form a blanket 1 to 100 feet deep (average 50 ft) where the uppermost layers are very lean in nickel (< 1% Ni) and the rich ore zone occurs near the base of the weathered zone. The average content of Ni in unserpentinized dunites is 0.30% and for harzburgites 0.29% (data from section on peridotites). Olivine in peridotites contains 0.21 to 0.31% Ni (Nishimura et al., 1968) and is the main source of the nickel in the peridotites, however, Guillon and Lawrence (1973) report accessory nickel sulfides (pentlandite, millerite, heazlewoodite) in the peridotites from New Caledonia. It is generally believed that the nickel concentrated in the laterite ores is derived from the breakdown of olivine and pyroxenes during weathering (De Vletter, 1955; Trescases, 1969; Hotz, 1964). However, a small proportion of the nickel may have been derived from Ni released from the nickel sulfide accessory minerals in the peridotite (Guillon and Laurence, 1973).

The concentration of nickel within the weathered zones of the peridotites is regulated by dissolution of olivine and orthopyroxene in the peridotites and removal of magnesium and silica along with residual concentration of iron and nickel (Fig. 53). Nickel is then dissolved in the uppermost layers of the laterite and becomes enriched downward in the lower parts of the weathered zone (Chetelat, 1947; De Vletter, 1955; Hotz, 1964). Assuming that no nickel was lost during the weathering processes, a peridotite containing 0.25% Ni and yielding an average ore of 1.5% Ni indicates 1:6 concentration and an ore of 3.5% Ni would indicate 1:14 concentration. Similar concentration ratios can be calculated for the iron enrichment (1:6, 1:8) but here the higher downward enrichment factors seen for the nickel do not obtain for the iron deposits.

4.2 Asbestos

During the serpentinization of the peridotites from ophiolites fibrous chrysotile is formed as cross fiber veins or agglomerates of finely matted material. Where the fibrous chrysotile forms high enough concentrations it is mined as asbestos which has many uses in industry because of its fibrous structure, high heat resistance, non-corrosion by chemicals, and ability to act as a fireproof thermal insulator. There are many variations in the fibrous structure and color which is manifested by ease of fiber separation and silkiness of the fibers. Each deposit shows wide variations in content and chrysotile structures. It is estimated that 4.8 million tons of asbestos are produced in the world annually and of that at least 50% comes from deposits developed in ophiolite peridotites.

The most important area of asbestos production is located in the Appalachian belt of Paleozoic ophiolites extending through southern Quebec, Canada and northern Vermont, U.S.A. (Lamarche, 1972; Laurent, 1975). The ophiolites of southern Quebec are thrust sheets that have been emplaced during early Ordovician time and lie above the Cambrian metasediments of the Caldwell Group (Lamarche, 1972; Laurent, 1975). Overlying the ophiolites is a mélange considered to be Early Ordovician in age (Laurent, 1975). The ophiolite occurs as stratified sheets with a thick basal peridotite and thin upper unit of gabbro and volcanics (Fig. 54). The main ore zone for asbestos is contained in the basal harzburgite (Riordon, 1954). The present mining activity is concentrated in seven open pits and two underground mines and the ore consists of both cross fiber and slip fiber. The most important ore is a stockwork of cross fiber veins that average 6 mm but may attain widths of 25 mm. Shearing and faulting produce slip-fiber seams (Cooke, 1937; Riordon, 1954; Riordon and Laliberte, 1972). Formation of the asbestos deposits seems to be linked to early fracturing and faulting of the serpentinized harzburgite and intrusion of leucocratic dikes into the peridotite after fracturing. Circulation of hydrothermal fluids within the harzburgite allowed development of the spectacular cross fibre ore

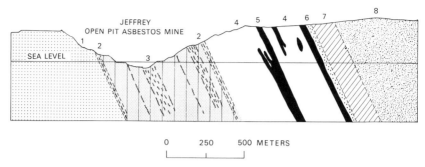

Fig. 54. Cross section through the asbestos-bearing ophiolite, Jeffrey mine, Johns-Manville Co. Ltd., Quebec, Canada. *1* Cambrian Caldwell Group (country rock); *2* Sheared serpentinite; *3* Alpine-Peridotite (asbestos-bearing harzburgite); *4* dunite; *5* pyroxenite; *6* gabbro; *7* volcanics and breccias; *8* Early Ordovician, St. Daniel Formation (mélange). (Modified after Laurent, 1975)

(Cooke, 1937; Riordon and Laliberte, 1972). There are of course many peridotites associated with ophiolites that have had somewhat similar tectonic histories but do not contain economic deposits of asbestos.

Another type of asbestos deposit found within the serpentinized ophiolite peridotites may form when the tectonic forces completely dismember the ophiolite and the serpentinite develops into a highly sheared mélange or diapirically moves upward during subsequent orogenic events. In most cross- or slip-fiber veining asbestos deposits only 5–10% of the ore is useable whereas in the tectonized serpentinized bodies there may be as high as 50% recoverable chrysotile. However, in these deposits only a short fiber asbestos (Grade 7, Canadian fiber) is produced (Mumpton and Thompson, 1975).

The Coalinga asbestos deposit in western California has one of thre greatest reserves of this type of asbestos ore in the world and its annual production represents 1.5% (75,000 tons) of the total annual world's asbestos production. The ore from the Coalinga deposit consists of a soft powdery, pellet-like agglomerates of finely matted chrysotile $(<2\,\mu)$ that forms a matrix containing blocks and fragments of solid serpentine that may range in size from a few mm up to several meters (Mumpton and Thompson, 1975). In fact, the ore could really be characterized as the matrix to a serpentinite mélange. The tectonized and soft nature of the ore allows open-pit mining and processing is by both wet and dry milling processes (Merritt, 1962; Reim and Munro, 1962; Mumpton and Thompson, 1975). The Coalinga asbestos deposits occur within the New Idria serpentinite mass which is 6.4×19.2 km and occupies the crest of the Coalinga antiform. Gravity surveys of the mass suggest that it extends downward nearly 4500 m (Byerly, 1966) and the geologic situation requires upward diapiric movement of the serpentinite starting in early Miocene and continuing up until the present (Coleman, 1961). The position of the New Idria serpentinite within the Franciscan and Great Valley sediments suggests that it was once part of the ophiolite complex that forms the basement for the Great Valley and/or Franciscan sediments (Bailey et al., 1970) and that it reached its present position by repeated tectonic movements by becoming a very light ($\sim$ S.G. 2.52) rock mass that moved as a vertical diapir. Formation of the asbestos was a result of this continued shearing and pulverization accompanied by dissolution and re-precipitation of chrysotile along sheared surface (Barnes and O'Neil, 1969; Mumpton and Thompson, 1975).

The formation of asbestos within the peridotites from ophiolites must have diverse origins. It is clear from the above discussion that the asbestos is related to the process of serpentinization, but it is not clear what chemical or physical parameters favor the formation of the asbestos ores.

Part VI. Geologic Character

1. Ancient Convergent Plate Margins (Sutures)

Throughout the world, ophiolites are exposed typically along belts of intense tectonism (Fig. 55). They occur along major geosutures and are thought to mark the sites of ancient zones of interaction between oceanic and continental crust, even though many parts of the older belts are now well within the interior of some continental areas. Thus, although ophiolites are generated at accreting plate margins they are either subducted and lost—or else obducted and preserved—along convergent plate margins.

In general, structures within the suture zones parallel the trend of the zones. Mélanges comprising parts of ophiolite mixed with enclosing sedimentary and metamorphic rocks commonly mark the boundaries of the suture zones (Gansser, 1974). The common occurrence of blueschist metamorphic belts in suture zones containing ophiolites is another significant feature (Coleman, 1972b). Amphibolite and green-schist facies rocks are also present in some areas, but most of the metamorphic rocks of these suture zones seem to have formed as a result of regional dynamothermal conditions rather than by contact meta-morphism, but their deformational style seems to be related in some way to the tectonic emplacement of the ophiolites.

The existence of ophiolite within the older Precambrian cratons has not yet been satisfactorily resolved. Naldrett (1972) has concluded that in the ancient Archean greenstone belts there are gravity-stratified sills and ultramafic lenses emplaced with eugeosynclinal sediments penecontemporaneous with volcanism. In some areas there is evidence that at least some of the ultramafic lenses represent extrusive ultramafic lava, see for instance Naldrett and Mason (1968), Viljoen and Viljoen (1969), Nesbitt (1971) and Glikson (1971). These Archean greenstone belts are considered to be autocthonous masses and to have formed under conditions unique to the Archean. Future work will undoubtedly resolve the problem of the Archean greenstones vs. Phanerozoic ophio-lites and for the purposes of this discussion, these ancient greenstones will be excluded.

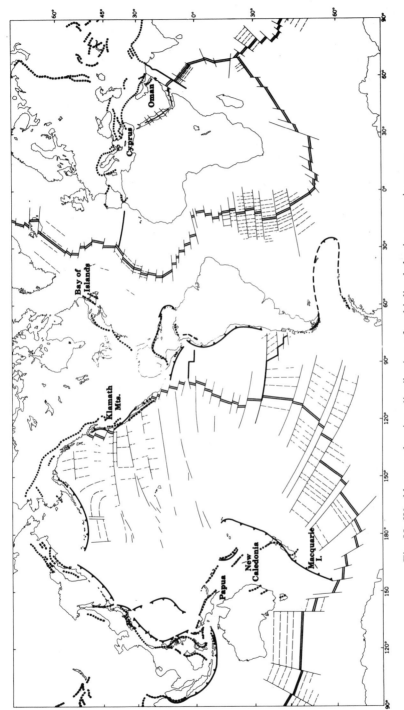

Fig. 55. World map showing distribution of ophiolite belts in orogenic zones

Early Paleozoic ophiolites are represented by the Appalachian and Caledonian belts of eastern North America, western Fennoscandia, and the Grampian Highlands of England. The Bay of Islands ophiolite is the best known of these early Paleozoic ophiolite masses and will be described in detail (see Part VII). Perhaps the largest concentration of Paleozoic ophiolites is in the Ural Mountains which continue South and then eastward into Mongolia, and include central Kazakhstan, Tien Shan, Altai-Sayan, western Siberia, and the Mongolia-Okhotsk folded region (Burtman et al., 1973). Along the Pacific Coast of North America, Paleozoic ophiolites are present in the Klamath Mountains of Northern California, in western Canada, and in northern Alaska. In eastern Australia, Paleozoic ophiolites extend from Quensland southward into Tasmania.

Mesozoic and Cenozoic ophiolites are much more abundant and constitute most of the important large exposures of ophiolite. The Tethyan belt extends from the Betic Cordillera and Rif of Spain and Africa eastward through the Alps, the Dinarides in Yugoslavia, through Greece, Turkey, Iran, Oman, Pakistan, the Himalayas, Burma, and Indonesia, where it connects with the Circum-Pacific ophiolite belt. Within the Pacific area, ophiolite belts of Mesozoic and Cenozoic age extend from New Zealand northwesterly to New Caledonia, New Guinea, Celebes, Borneo, Philippine Islands, Japan, Sakhalin, Kamchatka, and the Koryak-Chukotka region, and from Alaska southward along the western Cordillera of North America. Eastward from Guatemala, the ophiolite trends through the Greater Antilles, Cuba, and Puerto Rico and loops into the Caribbean Andes of Venezuela southward through western Colombia and terminating in Ecuador.

It is beyond the scope of this book to carefully document all of the occurrence of ophiolites enumerated above. At the present time, the literature on ophiolites is expanding so rapidly it is virutally impossible to summarize new information on ophiolite occurrences. A recent symposium in Moscow brought together many of the world's leading experts working on the various aspects of the ophiolite problem (Coleman, 1973) and the abstracts of the papers presented at this meeting provide an excellent summary of the tectonic and geologic setting of ophiolite belts throughout the world (Anonymous, 1973) as known up until that time.

2. Associated Sediments

Because the ophiolite complexes now appear to represent alloch- thonous transported fragments of oceanic litosphere, the sediments in contact with them can provide vital evidence regarding the origin of

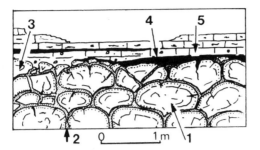

Fig. 56. A section showing the relations between pillow lavas and associated pelagic sediments. *1* A complete pillow lava with its variolitic border; *2* hyaloclastic sand; *3* a fragment of pillow, *4* radiolarian chert; *5* calcilutite and detrital diabase grains included. In this case, the pillows are obviously in primary association with the sediments. (After de Graciansky, 1973)

the ophiolite as well as something about its age and possible modes of transport. Assuming that the ophiolite represents a fragment of oceanic lithosphere transported onto a continental margin, it should retain certain geologic evidence of its igneous origin within the oceanic lithosphere as well as stratigraphic indicators revealing its age of tectonic emplacement. Advances in our knowledge of the deep ocean sediments deposited on oceanic crust have provided us with new information regarding the type of materials resting directly on the oceanic crust and interlayered with the pillow lavas (Thurston, 1972; Robertson and Hudson, 1973; Berger, 1974; Robertson, 1975). There are present pelagic sediments and chemical precipitates resting directly on or interbedded with the pillow lavas that form the top portion of ophiolite complexes (Abbate et al., 1972; Mesorian et al., Robertson and Hudson, 1973; Glennie et al., 1974) (Fig. 56).

Generally cherts and shales are the most common pelagic sediments and these contain organic remains that clearly demonstrate their deep ocean origin; Steinmann, (1927) emphasized this in his early papers (Fig. 57). Where these sediments clearly are interbedded with the volcanic materials, minimum ages can be estimated by identification of the microfossils. The presence of chemical precipitates (umbers) within the volcanics, as shown earlier, is considered to be part of the chemical processes related to hydrothermal alteration of the oceanic crust developing at spreading ridges (Spooner and Fyfe, 1973; Robertson and Hudson, 1973; Spooner, 1974). These same types of iron-rich chemical sediments are now forming on the East Pacific Rise (Corliss, 1971). Thus, the occurrence of pelagic sediments and umbers within the upper parts of ophiolites provides geologic evidence of a deep ocean

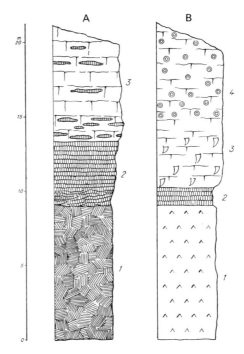

Fig. 57A and B. Sequences on top of ophiolites SW of Konya, Turkey. (A) South-west of Hatip. *1* serpentine; *2* radiolarian cherts (disrupted at the serpentine contact); *3* micritic limestone with chert lenses. *Globochaete alpina*, pelagic pelecypods, ostracods, sponge spicules. Dogger? upper Jurassic. (B) North of Cegir Bagi. *1* spilitic lavas; *2* radiolarian cherts; *3* limestone with Rudistae and algae, Upper Cretaceous; *4* oolitic limestone with small gastropods and algae, Upper Cretaceous. The two outcrops are within a few kilometers; their original distance is unknown, since they are part of an ophiolitic melange. (After Passerini and Sguazzoni, 1966, and shown in Abbate et al., 1972)

history and hydrothermal activity related to submarine volcanism. In many cases, rapid erosion combined with weathering may destroy the pelagic sediments capping an ophiolite slab. Sharp unconformities or disconformities marking changes from pelagic to shallow water or trench deposits within sediments deposited on top of the ophiolite indicate the end of tectonic transport from an oceanic basin to a continental margin, and if fossils are present a minimum age for tectonic emplacement can be established, (see for instance Abbate et al., 1972 and Glennie et al., 1974). Sediments underlying the ophiolite may also be allochthonous slabs of pelagic sediments or trench sediments transported during tectonic movement of the oceanic crust and may contain fossils of the same age as those interlayered with the volcanics.

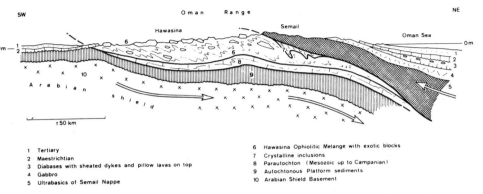

1 Tertiary
2 Maestrichtian
3 Diabases with sheated dykes and pillow lavas on top
4 Gabbro
5 Ultrabasics of Semail Nappe

6 Hawasina Ophiolitic Melange with exotic blocks
7 Crystalline inclusions
8 Parautochton (Mesozoic up to Campanian)
9 Autochtonous Platform sediments
10 Arabian Shield Basement

Fig. 58. Schematic cross section through the Oman Semail nappe showing the
Hawasina zone. (After Gansser, 1974)

In other cases, the underlying sediments may be autochthonous and fix
a maximum age of ophiolite emplacement. Continued studies of
sediments associated with ophiolite complexes will provide direct
geologic evidence as to age of igneous formation and tectonic transport
as well as information bearing on the nature of the continental margin
at the time of its emplacement.

3. Mélanges

Many ophiolites occurring within folded mountain belts have undergone repeated deformation and metamorphism, which transforms the
layered fragments into tectonic mixtures or mélanges. Recent work has
revealed that the ophiolites are usually large allochthonous tectonic
slabs which are now parts of deep seated nappes in the Urals, Tien Shan,
Appalachians, Klamaths, and the Alps (Williams, 1971; Peive, 1973;
Coleman and Irwin, 1974; Dietrich et al., 1974; Gansser, 1974). Transformation of the peridotitic parts (dunite-harzburgite) into serpentinite
early in the tectonic movements of ophiolites produces a material
whose response to tectonic transport is somewhat analogous to that of
salt. Sheared serpentinites are less dense than the rocks surrounding
them and commonly provide a plastic medium into which the mafic
fragments of an ophiolite, associated sediments and sometimes metamorphic rocks become incorporated (Mercier and Vergely, 1972). Once a
mélange with predominately serpentinite matrix is developed, it is
vulnerable to any future tectonic event. Because the sheared serpentinite
matrix undergoes plastic deformation under stress (Coleman, 1971b),
a mélange with a serpentinite matrix responds to tectonism by

movement, complicating its inner structure and incorporating blocks of country rock through which is passes. Considerable confusion surrounds the significance of mélange and its relationship to the emplacement of ophiolite slabs (Gansser, 1974) (see Fig. 58). Where gravity sliding is important, the presence of mélange is considered to be part of a chaotic mixture of blocks produced by the effects of gliding down slope (olistostrome). In other situations the associated mélange has been interpreted as a remnant of an ancient subduction zone operative during the emplacement of the ophiolite (Bailey et al., 1970; Hsu, 1971; Blake et al., 1974; Cowan, 1974).

Part VII. Emplacement Tectonics

1. Introduction

Processes leading to the emplacement of ophiolite are most easily tied to plate motions rather than to in situ igneous intrusions. Interactions between plates are considered to give rise to orogeny and many of the younger orogenic belts can be related to zones of plate convergence (Fig. 59). Interactions within certain zones of convergence such as the Alpine-Tethyan orogenic belt, allochthonous ophiolite masses are imbricated with nappes whose origin is quite different from that of the ophiolite (Gansser, 1974). Formation of ophiolite at spreading ridges, marginal basins, or rifts requires that these fragments of oceanic crust have been transported and incorporated into an orogenic zone. The major plate boundaries recognized are: (1) accreting; (2) consuming; (3) transform. For the purposes of this discussion, it is assumed that ophiolite is formed at accreting boundaries (spreading ridges) and that only a small fraction of the new crust formed at the spreading ridge is ever tectonically emplaced along a consuming or transform plate boundary (continental margin). Reconstruction of spreading centers in relationship to consuming margins during the Phanerozoic requires that most of the oceanic crust developed at the accreting boundaries has been consumed by subduction and destroyed in the asthenosphere. The amount of oceanic crust incorporated into the orogens of continental margins is an extremely small percentage ($<0.001\%$) of the total oceanic crust formed and thus it is apparent that tectonic emplacement relates somewhat to a major perturbation of the plate motions. The following discussion will attempt to categorize possible tectonic perturbations that could lead to the incorporation of ophiolite at "convergent" or "consuming" margins.

2. Obduction — Subduction

Steady-state subduction is required to consume the large amounts of oceanic crust developed at the spreading centers. The consumption is visualized as a bending of the oceanic plate downward and its sinking

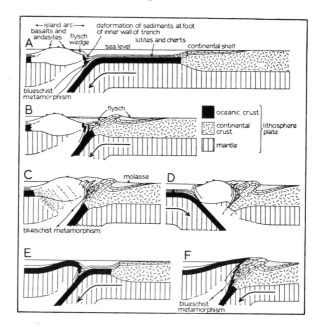

Fig. 59. (A–D) Schematic sequence of sections illustrating the collision of a continental margin of Atlantic type with an island arc, followed by change in the direction of plate descent. (E–F) Proposed mechanism for thrusting oceanic crust and mantle onto continental crust. (After Dewey and Bird, 1970, p. 2641, Fig. 12)

into the mantle where it is assimilated and reincorporated into the mantle (Oliver et al., 1969; Dickinson, 1971a, b; Turcotte and Oxburgh, 1972). However, steady state subduction does not allow parts of the oceanic crust to become detached and incorporated into the edge of the continent (Coleman, 1971a). It is visualized that pelagic sediments resting on top of the oceanic crust are not consolidated enough to withstand deformation and may be scraped off and incorporated into the trench sediments by a series of underthrust faults whose movements are synthetic to the subduction zone (Roeder, 1973; Helwig and Hall, 1974; Oxburgh, 1974). However, none of these synthetic faults appears to penetrate into the rigid oceanic crust even though this has been suggested as an ophiolite emplacement mechanism by Rod (1974), for the Papuan ophiolite (Davies, 1971). Ernst (1970, 1973, 1974) has proposed that both sediments and oceanic crust can be subducted as part of the downgoing slab to be metamorphosed under high P-low T conditions and later exhumed on the orogen. This requires return to the Earth's surface within the convergence zone and could only be accomplished by a cessation of steady-state subduction. The occurrence of blueschist assemblages within the Zermatt-Sass ophiolite of the Western Alps provides evidence that some parts of the subducted oceanic crust can be resurrected at convergent margins (Bearth, 1967; Dal Piaz, 1974); however, the tectonic conditions involving transport of metamorphosed slabs from such great depths remains obscure.

Obviously, tectonic processes related to steady state subduction cannot provide conditions that will allow detachment of ophiolite slabs up to 12 km thick from the downgoing oceanic lithospheric plate. The presence of large, unmetamorphosed ophiolite slabs overthrust onto the continental margins, however, provides direct geologic evidence that at least some of the oceanic crust had escaped subduction. To provide a tectonic term that would adequately describe this process and also mark it as distinct from the more common term "subduction," I introduced the term "obduction" (Coleman, 1971a). Obduction implies overthrusting at consuming plate margins but there may be numerous tectonic situations that would allow the detachment of oceanic crust prior to overthrusting (Fig. 60). Generally, it is considered that the thickness of the oceanic plates are 60–100 km (Oxburgh, 1974) and up to now the thickest known obduced ophiolite slab (12 km) is in Papua (Davies, 1971). The emplacement of such thin ophiolite slabs requires some sort of detachment surface to develop within the top portion of the oceanic plates. Armstrong and Dick (1974) have suggested that a steep geothermal gradient underlying the detached slab is necessary and describe the detachment as follows: "At the time of detachment, the relatively brittle and rigid cover moves away from its thermally softened base. At first, displacement between cover and base is penetrative, but as strain softening (and shear heating?) proceeds, displacement becomes localized along a fault zone that will be parallel to isotherms in the rock, and the overthrust crystalline-based sheet is thus freed from its base. During movement, the underlying rocks of the detached sheet may be metamorphosed, and fragments of the overridden rocks may be picked up." Situations that could provide the high heat flow are more likely to occur in rear-arc marginal basins or in small restricted ocean basins rather than where cold, thick slabs of oceanic crust are being subducted (Dewey, 1974).

There are numerous possibilities where young hot oceanic crust could be detached and obduced. Christensen and Salisbury (1975, p. 78) have suggested: "The youth of the ophiolite imposes a considerable restraint on possible mechanisms of emplacement. It is proposed that the most likely mechanism consistent with this phenomenon is the one shown in Figure 21 (see Fig. 61). During the closure of any ocean basin through subduction of one or both of its limbs the ridge crest itself must at some point be subducted. This point is unique in that for the first

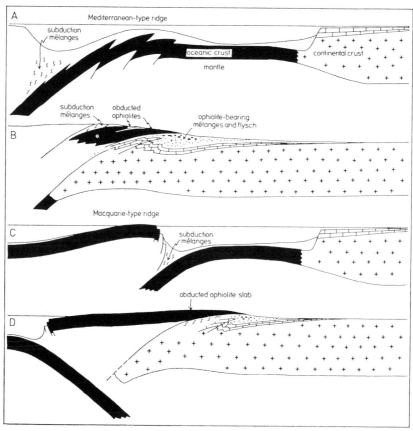

Fig. 60A–D. Possible mechanism for the obduction of ophiolite sheets onto continental margins. (After Dewey and Bird, 1971, p. 3193, Fig. 6)

and only time a thin hot mechanically weak segment of oceanic crust
and upper mantle, laced with magma chambers, is presented to the
subduction mechanism. That subduction of the ridge crest occurs
without incident is unlikely. It is anticipated, rather, that the ridge crest
will be dismembered by faulting, major segments, particularly from the
upper levels of the outboard plate, being obducted onto the continental
margin while the inboard plate is depressed under the approaching
continental plate and subducted."

Karig, (1972) has suggested that the polarity of an active arc system,
i.e., direction of subduction, may change and that this would lead to
arc–arc or continental margin remnant–arc collisions. In this situation,
high heat flow could be expected and cause shallow detachment of the

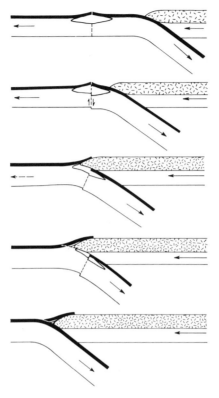

Fig. 61. Ophiolite emplacement during subduction of a ridge crest. (After Christensen and Salisbury, 1975, p. 79)

oceanic crust. The New Britain arc system is considered as the boundary between the Australian and Pacific plates and Karig (1972) regards the Papua ophiolite as a slice of marginal basin oceanic crust obducted during advancement of the Pacific plate towards the Australian plate.

Destruction of marginal basins by a change in the polarity of the associated arc systems combined with continental margin collision provides a hypothetical mechanism of emplacement for many ophiolite occurrences described from the Circum-Pacific area and also explains the close association of ophiolite with island arc assemblages. Considering present day plate tectonics, obduction is not observed and so documentation is dependent on geologic evidence. Dewey and Bird (1971) have also recognized a lack of obduction across active trenches in which oceanic crust is being transported from the subducted plate across the trench onto the upper surface of the continent. The lack of observed

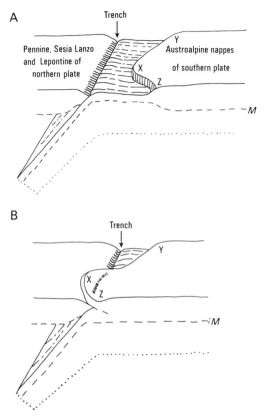

Fig. 62. (A) Schematic representation of an irregular continental margin *(ZXY)* approaching a trench at a continental margin. (B) Formation of a crustal flake after continental collision (═══), oceanic crust; − − − M-discontinuity; .... base of the lithosphere. (After Oxburgh, 1972, p. 204)

obduction, combined with apparent irregular emplacement of ophiolites throughout Phanerozoic time, requires that there be other mechanisms for ophiolite emplacement. Oxburgh (1972) has introduced the term "flake tectonics" for the collision of two continental crust-capped plates and describes the process as follows: "When a continental area approaches a subduction zone through the destruction of intervening oceanic lithosphere, one part of the continent will arrive at the subduction zone before adjacent areas (Fig. 62 a, b) unless there is a remarkable coincidence of shapes. Body forces at x should then resist further subduction, whereas elsewhere on the plate margin (y and z, for example), body forces favour continued subduction. Insofar as x, y, and z are all part of the same coherent plate, some kind of mechanical compromise must be reached at x until contact is made at sufficient points between continents for the plate boundary to become locked and subduction to cease.

"In the case of the Eastern Alps, the mechanical compromise involved splitting of the lithosphere within the continental crust. This splitting may have been partly associated with the arching of the lithosphere which occurs as the downturn into the trench is approached. Because the continental crust stands more than 5 km higher than oceanic crust, however mechanical locking could occur as the continental crust enters the subduction zone. I suggest that a low angle crustal split propagated backwards from this locking point and that flake separation then occurred (Fig. 62 b). A separation of the upper third of the continental crust would have the effect of reducing the total buoyancy of the subducted crust by nearly a half." Even though Oxburgh specifically excludes oceanic crust or mantle from his model, it seems unlikely that the detachment zone will always mark the interface between continental material and oceanic crust (mantle).

The work on the Lanzo lherzolite massif demonstrates that it is the bottom part of a "flake" formed along the Alpine suture (Ernst, 1973) a zone of convergence between two continental plates (Berckhemer, 1968; 1969; Nicolas et al., 1972). Thus, it is possible by continental collision to emplace heavy mantle material or oceanic crust as "flakes" over continental crust. Roeder (1973, Fig. 1) describes "flipped" subduction zones where change in the direction of subduction develops alpine root zones, rootless rotation zones, and overridden rotation zones and parts of the oceanic crust within older frozen subduction zone may be tectonically transported upwards into the orogen during changes in convergence polarity. Dewey (1974) and Dewey and Bird (1971) also prefer emplacement of ophiolites by obduction, but believe that the process represents the collision of continents or arcs and continents and not an abberation of steady state subduction. Dewey (1974; Fig. 7) illustrates the emplacement of the Appalachian-Caldonian ophiolites as a complicated series of openings and closing of marginal basins landward of a westdipping subduction zone. He visualizes the ophiolite occurrences as developing in separate rear-arc, and interarc oceanic basins during the Ordovician and being obducted shortly thereafter by the closing of these oceanic basins.

Abbate et al. (1973) have provided compelling evidence that all the Tethyan ophiolites were emplaced nearly simultaneously during a Late Cretaceous gravity sliding or thrusting (obduction). They also point out that spreading of the Tethyan sea had virtually stopped by Late Cretaceous and that the closing of the Tethyan sea was not accompanied by subduction (lack of volcanism). It is their opinion that the emplacement was due to convergence of small unconnected ocean basins. This leads to another variant of ophiolite emplacement which can be related to formation of small ocean basins such as the Red Sea. These basins would have only limited size, but would maintain a high heat flow until a later compressional event closed the basin with concomitant detachment of the upper parts of the oceanic crust. Obduction of the detached oceanic crust during closing of the small ocean basins would produce ophiolite allochthons resting on the basin edge sediments and capped by pelagic sediments of the basin axis.

3. Diapirs

There are a number of situations where the occurrence of ophiolite apparently requires an emplacement mechanism unrelated to plate motions. Maxwell (1970, 1973, 1974a, b) has maintained that diapiric uprise of hot mantle material through continental and oceanic crust best explains the internal location of these rocks in the orogen and describes this as follows (Maxwell, 1973): "Figure 63 illustrates a concept of diapiric formation of ophiolites where the ophiolite has breached the crust and differentiation products have accumulated above the ultramafic core.... The model provides for the formation of pillow lavas and extrusive breccias on the surface of an accumulating reservoir of basaltic lava, much of which subsequently cooled with a diabasic texture. During the process of growth of the diapir, continuous addition of basaltic material would expand the carapace and could logically give rise to sheeted dikes of the type characteristic of Cyprus and other large ophiolite slabs.... A mushrooming of the upper part of the diapir might logically result in the emplacement of hot ultramafic rocks and overlying diabases and lavas upon volcanic or sedimentary rocks, producing the occasionally observed contact metamorphism. The model also explains the very commonly observed characteristic of ophiolites that the thinner sheets usually lack gabbro, and that locally, diabase, pillow lavas or even radiolarian cherts may lie directly on serpentinite." Maxwell's model is an extension of the submarine extrusion hypothesis of Brunn (1961) and provides a means of overcoming the large lateral transport of oceanic crust across subduction zones at continental margins required by obduction emplacement. Furthermore, the production of small batches of oceanic crust by this model does not later require consumption of new oceanic crust formed at a spreading center.

Chidester and Cady (1972) have provided evidence that the small isolated peridotites in the Appalachian belt were emplaced diapirically

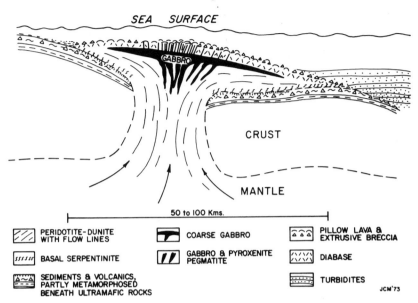

Fig. 63. Cross section of hypothetical ophiolite diapir. (After Maxwell, 1970, 1973)

through continental crust: "These conditions led to a complex history of tectonic emplacement and metamorphism. Kneading of the solid periodite upward through sialic crust and eugeosynclinal sediments led to their separation from the less mobile components of the ophiolitic complexes; their closest associates are the basaltic volcanics intercalated in the eugenosynclinal assemblage, which corresponds to the sheeted diabase complexes and pillow lavas of the ophiolite assemblages."

Diapiric uprise of material within the mantle and crust usually requires that the rising material have a density less than its surrounding material and a plasticity that would allow such migration. For the peridotites from ophiolites, this would entail either a magma or transformation of the peridotites to serpentinite. The paucity of contact aureoles around the boundaries of ophiolites has been the most damaging evidence against hot diapir emplacement. On the other hand, diapiric uprise of serpentinites has been documented in numerous instances (Coleman, 1971 b, p. 905). Loomis (1972 a, b, 1975) has provided evidence that the Ronda and Beni Bouchera high temperature peridotites on opposite sides of the western Alboran Sea were emplaced as hot diapirs. The presence of extensive contact metamorphic aureoles combined with geophysical (Bonini et al., 1973) evidence that shows these bodies were rooted in the mantle and is new data that overcomes some objections to hot diapiric emplacement. The Ronda and Beni Bouchera consist primarily of lherzolites with metamorphic textures but are not associated with the typical gabbros, diabase, or basalts of ophiolite and therefore, they probably represent a continental mantle rather than oceanic lithosphere (Nicolas and Jackson, 1972). Loomis (1975) describes the Middle Tertiary emplacement of Ronda and Beni Bouchera ultramafic masses as part of crustal thinning and mantle upwelling due to extensional regional strain produced by rotation of the Iberian Peninsula relative to Africa.

4. Gravity Slides, Protrusions, Deep Faults

There have been numerous other mechanisms invoked to explain the emplacement of ophiolite, all of which may be valid for certain situations but do not appear to have a universal application. It is easy to conceive a complicated history for an ophiolite after it becomes part of the continental crust, particularly if the peridotite member of the ophiolite becomes serpentinized and incorporated into a mélange.

The tectonic evolution of the northern Apennines has been carefully worked out by numerous authors (Decandia and Elter, 1969; Abbate et al., 1970; Elter and Trevisan, 1973) and here ophiolite emplacement by gravity sliding has been documented. Oceanic crust formed during Jurassic times is covered by radiolarites and overlying Calpionella limestone and is referred to as the Ligurian domain. Uplift of the Ligurian oceanic crust with its carapace of Jurassic cherts and pelagic limestones produces the Bracco ridge giving rise to widespread gravity sliding where large ophiolite blocks (several cubic km) were transported on lenses of breccia (olistostrome). This event was followed by over-thrusting (gravity sliding?) of the Ligurian olistostrome containing coherent blocks of ophiolite over continental and marine sediments of the Tuscan and Umbrian depositional basins during the Tertiary. Formation of serpentinite from the Ligurian peridotites facilitated the gravity sliding of these units.

Glennie et al., (1974) and Stonely (1975) visualize a similar mechanism for the emplacement of the Semail ophiolite. However, neither ex-planation provides a satisfactory tectonic mechanism for elevating the ancient oceanic crust high enough to initiate gravity sliding. There is no geodynamic reason for a deep ocean basin to rise isostatically unless there is initiated an upward diapiric movement within the mantle. On the other hand, the closing of a small ocean basin by plate movement could easily elevate of the oceanic crust during collision of two conti-nental plates by upwarping and thus initiate gravity sliding of ophiolite onto a continental margin. Thus, detachment of ophiolite from the upper parts of the oceanic lithosphere and concomitant gravity sliding most certainly is a viable emplacement process. Even though actualistic models of gravity sliding of oceanic crust away from mid-oceanic ridges can not be verified, it seems feasible to entertain this as a possible mechanism, particularly where heat flow is high and the oceanic lithosphere still retains some of its plasticity at shallower depths, (Bottinga and Allegre, 1973).

Numerous workers (Milovanovic and Karamata, 1960; Knipper, 1965; Lockwood, 1971, 1972) have been impressed by the tectonic mobility of serpentine once it forms and have referred to this phenomena as cold intrusion, solid intrusion, or tectonic intrusion. Lockwood (1972) has suggested the term *protrusion* to cover this tectonic process and provides us with the following model of emplacement: "According to this model, oceanic lithosphere, consisting at least in part of serpentinite, is subducted beneath the continental rise during early phases of geosynclinal development and slowly rises in temperature as it descends. At temperatures as low as 300° C, partial dehydration of serpentinite may begin. As described in the preceding section, dehydration of even minor amounts of serpentinite will produce local water pore pressure that cause rock fracturing and abrupt lowering of rock strength. At this point, an unstable situation exists wherein relatively low density, weak serpentinite is overlain by higher density rocks. This situation is akin to sedimentary sections containing salt, and like salt, serpentinite is susceptible to upward protrusion. The mechanisms which trigger protrusion are unknown, although in an area of subduction faults in the overiding plates are to be expected and may provide protrusion channels... once extruded on the earth's surface, serpentinite will flow downslope to form chaotic deposits (olistostromes) and exotic blocks (olistoliths) in a trench, in interrise basins, or on the deep sea floor." Examples of protruded serpentinites are extremely common and the mechanism proposed by Lockwood must obtain in numerous situations. Emplacement of the rather thick and only slightly deformed ophiolite slabs by this process, however, seems quite unlikely. The mobility of the serpentinized periodite within the orogen is perhaps the most important factor in the dismembering of an ophiolite and must always be considered in its tectonic reconstruction.

This brings us to consider emplacement of ophiolites along deep fundamental faults. Numerous Russian geologists call upon deep fundamental faults into the mantle to provide a tectonic situation to emplace solid peridotites within the earth's crust (Peive, 1945; Reverdatto et al., 1967; Khain and Muratov, 1969). The concept here is that the mobile orogenic zones are bounded by deep fundamental faults that persist for long geologic time. Most of the movement is considered to be vertical and that the fault extends through the earth's crust and

penetrates into the mantle. Association of discontinuous pods of ultramafic rock and high pressure metamorphic minerals along these deep
fundamental faults has been interpreted as signaling great vertical
movements whereby mantle material and high pressure metamorphic
rocks can be brought to the surface. Perhaps the best studied deep fault
is the Alpine suture which represents a fundamental structure in the
Alps. It seperates the Hercynian structures and metamorphism on the
south from the younger Alpine nappes to the north and was earlier
connected in with the concept of "root zones" (Gansser, 1968). Significantly, several peridotite bodies such as the Finero and Lanzo are
situated on the hanging wall of the Alpine suture. Geophysical studies
have shown that positive gravity anomalies follow the Alpine suture and
that seismic velocities indicate that exposed peridotite bodies may have
roots in the mantle (Berckhemer, 1968, 1969; Nicolas et al., 1971; Peselnick et al., 1974). The evidence indicates possible solid emplacement of
mantle peridotites along a deep fundamental fault. These peridotites
are, however, not typical of the ophiolite sequence and contain no
associated mafic rocks (Nicolas and Jackson, 1972). It therefore becomes
obvious that there are numerous possible ways to emplace ophiolites,
but confusion develops because numerous non-ophiolitic mafic-ultramafic rock bodies have been used to demonstrate emplacement modes
for ophiolites.

The high temperature lherzolites such as Beni Bouchera, Ronda,
Lanzo, Finero are clearly different than the ophiolites and should be
treated as a special class as suggested by Nicolas and Jackson (1972).
The deep fault and mantle diapir emplacement of the high temperature
lherzolites seems required to preserve their high temperature-pressure
assemblages, whereas thin skin tectonics such as gravity sliding, initiated
by subduction or obduction can best explain emplacement of most
ophiolite slabs. Postemplacement deformation of ophiolite slabs combined with serpentinization and metamorphism, demand careful geologic studies before the actual sequence of emplacement events can be
sorted out. The complications arising from metamorphism (serpentinization) are treated in the chapter on metamorphic petrology.

Part VIII. Geologic, Tectonic, and Petrologic Nature of Four Ophiolites

1. Introduction

Four ophiolites from various parts of the world were selected for integrated descriptions of their petrologic, tectonic, and geophysical nature. This chapter provides essential data for evaluation of these ophiolites as possible candidates that may represent ancient oceanic lithosphere. I have either visited or carried out work on three of these occurrences and have had numerous discussions with Dr. Hugh Davies on the Papua-New Guinea ophiolite, and with Professor H. Williams on the Bay of Islands ophiolite. Thus, the descriptions that follow have the advantage of a single viewpoint with the aim of integrating the information into a rather uniformitarian presentation. Conceivably other workers with different viewpoints would choose to emphasize certain aspects that I have neglected; nonetheless, the basic data are presented with a minimum of interpretation.

It soon becomes apparent that there is a distinct lack of careful petrologic studies on these four ophiolites. Future comparisons of these sequences with the more abundant petrologic data from oceanic igneous rocks almost certainly demands that the geologic community direct more effort towards the study of ophiolites. This future work should concentrate not only on the petrologic problems but also include careful stratigraphic, structural, metamorphic, and geophysical studies of each occurrence to allow a proper synthesis to be made.

The close relationship between ophiolites and major geosutures has projected these rocks into prominence as indicators of ancient plate boundaries. The four examples described in this chapter appear to occupy such boundaries and are used as evidence to support these concepts. However, there is enough contrary discussion suggesting, that in some cases, these ophiolites may represent ancient island arcs and, if so, reasonable judgement should be exercised where less well studied mafic-ultramafic complexes are being considered as potential ophiolites marking ancient sutures. A number of aspects need to be clarified before an ophiolite can contribute evidence towards the solution of plate tectonics problems. The allochthonous nature can only be established by careful mapping and this aspect is well shown in the maps presented in this chapter. The igneous age of the rocks forming the ophiolite can establish its magmatic birth as can (interlayered) pelagic sediments within the pillow lavas. Metamorphic aureoles at the base of the ophiolites dated by radiometric methods can provide a maximum age of tectonic emplacement and transgressive younger sediments overlying the ophiolite will document a minimum age of emplacement. Geophysical measurements can provide structural evidence as to size and shape of the ophiolite. Petrologic relationships should indicate igneous processes that could characterize formation of the oceanic crust rather than a continental igneous province. The following descriptions attempt to cover as many of these aspects as possible in each section but it will soon become apparent to the reader that not all the data necessary to draw conclusions are available. Future investigations concerning the origin of ophiolites should concentrate on providing this information.

2. Bay of Islands Ophiolite, Newfoundland

2.1 Geologic Situation

In western Newfoundland, ophiolite complexes are present in the vicinity of Hare Bay and Bay of Islands as allochthonous slices forming the highest structural members of the Northwest Platform (Fig. 64). The Bay of Islands ophiolite complex forms a northeast-trending zone 96 km long and 24 km wide consisting of four separate mafic-ultramafic complexes (Fig. 64). There are numerous excellent reports on the Bay of Islands ophiolite complex and it perhaps is one area that has a nearly complete geologic documentation (Ingerson, 1935, 1937; Cooper, 1936; Buddington and Hess, 1937; Smith, 1958; Rodgers and Neale, 1963; Stevens, 1970; Church and Stevens, 1971; Dewey and Bird, 1971; Williams, 1971, 1973; Church, 1972; Williams and Malpas, 1972; Williams and Smyth, 1973; Dewey, 1974).

The whole allochthonous sequence rests upon an autochthon whose basement consists of Precambrian crystalline rocks of the Grenville province (800–1000 m.y.). The Grenville crystalline inlier of Precambrian is considered to be the ancient continental margin of eastern North America (Williams and Stevens, 1974). Resting unconformable on this basement is a Cambrian-Ordovician sedimentary sequence consisting of basal westerly derived clastics that grade upward into a thick carbonate sequence considered to have been deposited in shallow water

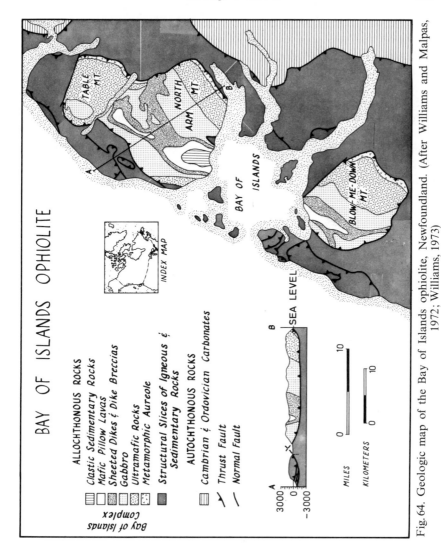

Fig. 64. Geologic map of the Bay of Islands ophiolite, Newfoundland. (After Williams and Malpas, 1972; Williams, 1973)

along the ancient margin of the North American continent (Humber Arm Slice assemblage). The transported structural slices that overlie the continental shelf carbonates consist of clastic sediments whose source was easterly and contributed debris from an exposed ophiolite (Stevens, 1970). This allochthonous Cambrian-Ordovician assemblages is equivalent in age to the underlying western platform sequence but was deposited in a trough that was to the east of the continental margin (Williams and Stevens, 1974). There is also evidence that mafic volcanic and intrusives invaded parts of this clastic wedge. The Bay of Islands ophiolite, the top slice of the allochthon, is locally underlain by mélange, metamorphosed mafic, and volcanic rocks and, on its eastern margin, a thin discontinuous amphibolite is welded to the bottom of the ophiolite slice. The age of the Bay of Islands volcanics is not known as there are apparently no fossiliferous sediments interlayered with the volcanics in the upper parts of the complex. However, the emplacement age or westward transport of the various slices has been estimated as pre-Middle Ordovician, as a transgressive limestone of Middle Ordovician rests unconformably on parts of the allochthon and exposed autochthonous rocks (Dewey, 1974). Concordant ages on zircons from trondhjemites associated with the Bay of Islands ophiolite indicate an igneous age of 508 m.y. (Mattinson, 1975) and K/Ar ages on the hornblendes from the metamorphic aureole give 460 m.y. (Dallmeyer and Williams, 1975).

It is now generally accepted that the Bay of Islands ophiolite represents a segment of the proto-Atlantic oceanic lithosphere generated at a spreading center or marginal basin (Church, 1972; Dewey, 1974; Williams and Stevens, 1974). Emplacement of the ophiolite occurred sometime during Early and Middle Ordovician as part of the closing of the proto-Atlantic Ocean. This event has been referred to as the Taconic orogeny within the Appalachians and has been aptly described by Williams and Stevens (1974, p. 793) as follows: "The mechanism of breakup is not clear but it resulted in the formation of a series of offshore island arcs, the obduction of oceanic lithosphere and mantle across the continental margin onto the shelf, the deformation and metamorphism of sediments bordering the continent, the mass transfer of continental-slope sequences westward across the shelf, and a flood of clastic sediments from the continental margin that transgressed toward the continental margin and almost complete closing of a proto-Atlantic Ocean in the northern Appalachians." Significant to this summary is the undoubted transported nature of the Bay of Islands ophiolite and complete absence of feeder dikes or necks within either the allochthon or autochton.

2.2 Internal Character

The Bay of Islands ophiolite complex can be generally divided into the following units which form a superposed sequence from bottom to the top: (1) Metamorphic aureole consisting of amphibolites apparently grading within a few meters downward into greenschist assemblages; (2) Peridotite section; (3) Transition Zone (critical zone of Smith (1958); (4) Gabbro section; (5) Sheeted dike complex; and (6) Volcanics (Fig. 65).

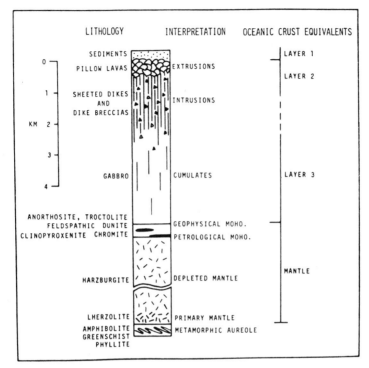

Fig. 65. Interpretation of the Bay of Islands ophiolite as oceanic crust and mantle. Various thickness estimated from geologic mapping. (After Williams and Stevens, 1974, Fig. 4)

Maps of Smith (1958) and Williams (1973) show that there is a strong disconformity between the peridotites and overlying transition-gabbro zone.

The peridotites (Table 15) have a distinct banded or layered structure that seems to have a regular trend over great distances. Williams and Stevens (1974) and Smith (1958) show generally steep (up to 70°) northeasterly dips in Table Mountain and North Arm Mountain. The bands or layers apparently are not isoclinally folded and so an estimated thickness of 4 km can be ascertained from the North Arm Mountain cross-section of Williams and Stevens (1974). The peridotite consists of two main rock types, dunite and harzburgite, and the observed banding results from variation in the amounts of olivine and orthopyroxene. Apparently the bands are discontinuous, pinching and swelling without developing a recognizable cumulate fabric (Smith, 1958). Parts of the peridotite section I examined along Trout River Pond exhibited strong

Table 15. Average chemical composition and CIPW norms of the various units within the Bay of Islands ophiolite, New Foundland[a]

	1	2	3	4	5	6	7	8	9
SiO_2	43.05	41.78	41.85	46.62	47.81	49.43	51.80	50.22	75.38
Al_2O_3	2.52	2.58	6.82	21.74	21.32	18.78	15.79	16.14	13.27
FeO	7.82	7.91	8.95	4.47	5.47	6.77	9.31	9.56	2.66
MgO	45.00	45.60	36.73	9.63	8.18	8.05	8.00	6.09	1.21
CaO	0.80	1.24	4.21	15.79	13.87	13.11	9.37	10.58	1.11
Na_2O	—	0.11	0.16	1.41	2.69	2.99	3.79	4.59	6.03
K_2O	—	—	—	—	—	0.21	0.53	0.64	0.11
TiO_2	—	—	0.28	0.15	0.52	0.62	1.05	1.82	0.17
P_2O_5	—	—	0.01	—	—	—	0.11	0.11	0.02
MnO	0.11	0.13	0.14	0.05	0.08	0.01	0.21	0.21	0.05
Cr_2O_3	0.39	0.33	0.60	0.11	0.04	0.02	0.03	0.02	—
NiO	0.31	0.33	0.25	0.03	0.02	0.01	0.01	0.01	—

Normative minerals

	1	2	3	4	5	6	7	8	9
Q	—	—	—	—	—	—	—	—	33.6
C	1.1	0.2	—	—	—	—	—	—	1.3
Or	—	—	—	—	—	1.2	3.1	3.8	0.7
Ab	—	1.0	1.3	8.2	15.5	20.0	32.1	25.8	51.0
An	4.0	6.1	18.0	53.0	46.1	37.2	24.5	21.5	5.3
Ne	—	—	—	2.0	3.9	2.9	—	7.1	—
Di	—	—	2.3	20.3	18.3	22.6	17.5	25.0	—
Hy	15.2	3.7	4.1	—	—	—	2.6	—	7.7
Ol	79.1	88.6	72.8	16.0	15.1	14.9	17.9	13.1	—
Cm	0.6	0.5	0.8	0.2	0.1	0.03	0.05	0.03	—
Il	—	—	0.5	0.3	1.0	1.2	2.0	3.4	0.3
Ap	—	—	0.03	—	—	—	0.3	0.3	0.05

[a] All analyses normalized to 100% after removal of H_2O, CO_2, and all iron recalculated as FeO.

(1) Peridotite section, av. 19, Table Mt. (Irvine and Findlay, 1972, Table 2). (2) Peridotite section, av. 13, North Arm Mt. (Irvine and Findlay, 1972, Table 3). (3) Feldspar dunite cumulate (Transition Zone), av. 5, Table Mt. (Irvine and Findlay, 1972, Table 2). (4) Olivine gabbro cumulate (Gabbro section), av. 20, Table Mt. (Irvine and Findlay, 1972, Table 2). (5) Olivine gabbro cumulate (Gabbro section), av. 10, North Arm Mt. (Irvine and Findlay, 1972, Table 3). (6) Gabbro, av. 2, North Arm Mt. (Williams and Malpas, 1972, Table 1). (7) Sheeted dikes, av. 6, Bay of Islands (Williams and Malpas, 1972, Table 1). (8) Pillow lava, av. 3, Bay of Islands (Williams and Malpas, 1972, Table 1). (9) Albite granite, North Arm Mt., (Irvine and Findlay, 1972, Table 3).

tectonite fabrics at the base and these seemed to persist with less intensity nearly to the transition zone. The dunite contains fractured and deformed grains of olivine; however, advanced serpentinization has consumed much of the primary silicates obscuring original textures. Harzburgite usually contains approximately 75% olivine and 25%

orthopyroxene and also has undergone intense serpentinization (Smith, 1958). Accessory chromite is present in both the dunite and harzburgite. Irvine and Findlay (1972) report a range of 0.04 to 0.31% Cr_2O_3 within the peridotite zone. The olivines from the peridotite section are remarkably constant in composition (Fo_{88-92}) and show no progressive vertical variation. This same uniformity is found with the $Mg/Mg+Fe$ ratio for analyses of the whole rock peridotites and the mineralogical nature of the chromites also remains nearly constant. Irvine and Findlay (1972) concluded from their chemical data that the peridotite did not form by fractional crystallization but in fact more likely was probably a residue of partial fusion. Church and Stevens (1971) and Church (1972) report high pressure mantle assemblages near the base of the peridotite unit within lherzolite-ariegites. They report the following minerals: olivine-orthopyroxene-clinopyroxene-spinel (lherzolite), clinopyroxene-orthopyroxene-spinel (ariegite), olivine-kaersutite-orthopyroxene-clino-pyroxene-ceylonite (kaersutite lherzolite), and kaersutite-clinopyroxene-garnet-ceylonite (garnetiferous amphibole ariegite). The spinels and clinopyroxenes are very aluminous, suggesting a mantle origin. These rocks are, however, restricted to a very narrow zone at the base of the peridotite section and have an intimate relationship with metamorphic aureole described by Williams and Smyth (1973).

The aureole rocks have pyroxene-garnet-amphibole-plagioclase assemblages and show polyphase deformation. Williams and Smyth (1973) record decreasing deformation and metamorphic grade (amphibolites to greenschist) structurally downward from the peridotite contact into unmetamorphosed supracrustal basic volcanics in distances less than 300 meters. They interpret these aureoles as having developed as contact dynamo-thermal recrystallization resulting from obduction and transport of hot oceanic crust and mantle in the early stages of the Taconic orogeny.

The transition zone (Table 15) is characterized by the first appearance of plagioclase within rocks that show cumulate textures. It is not clear from available descriptions if undoubted cumulate structures have been found in the peridotite section. The banded or layered rocks of the transition zone consist primarily of calcic plagioclase, clinopyroxene, olivine with small amounts of chromite, magnetite, and sulphide. The layers vary from several centimeters up to nearly 100 m in thickness and appear to have only limited horizontal continuity (Smith, 1958). Plagio-clase-bearing dunite, troctolites, anorthosites, and clinopyroxenites are the main rock types present in the transition zone but may be inter-layered with cumulate dunite and clinopyroxenite. Up to now there is no information on the presence or lack of cyclic units in the Bay of Islands complex as have been described from layered mafic-ultramafic igneous sequences (Jackson, 1971). The olivine in these layered units is more iron-rich (Fo_{83-89}) than that in the underlying peridotites (Irvine and Findlay, 1972). The plagioclase is usually very calcic, greater than An_{80}, and shows no reaction relations with the olivine. There are small discordant bands and pockets of clinopyroxenite within the base of the banded zone and in the upper parts of the peridotite. Deformation of the layers within the transition zone is common and, as pointed out by Smith (1958), the layers of the transition zone are discordant with the underlying peridotite banding.

The gabbro (Table 15) section overlies the transition zone and, in the upper part of the transition zone, gabbro interlayers are common. Above the transition zone the gabbros are generally massive and show only faint banding. These gabbros are as abundant as the peridotites and consist of varying proportions of plagioclase (An_{70-80}) and clino-pyroxene with minor amounts of olivine (Fo_{80}) and late forming magnetite. Serial samples of gabbro taken from North Arm Mountain show a progressive decrease in the $Mg/Mg+Fe$ ratio upward from the peridotite contact and Irvine and Findlay (1972) have interpreted this as indicating that the gabbros and layered rocks of the transition zone may have formed by fractional crystallization. Quartz diorites (plagiogranites) (Table 15) have been described within the gabbroic section and these rocks apparently represent small scale differentiation within the gabbro section.

A sheeted dike complex (Table 15) strongly brecciated in parts, situated structurally above the gabbro section and below the overlying volcanics, is described by Williams and Malpas (1972). The general strike of the dikes is northwesterly and they have variable steep dips to the southwest and northeast. The dikes extend downward into the gabbro and trend normal to the main diabase-gabbro contact which is northeasterly. As the contact between the gabbro and sheeted dike complex is approached, the dikes increase within the gabbro until finally no screens of gabbro exist and the sheeted dike complex consists of 100% dike chilled against one another. This relationship clearly indicates that the dike swarm postdates the gabbro. Brecciation of the dike swarm is extensive and Williams and Malpas (1972) suggest that this brecciation results from a late stage igneous gas action or fluidization. The dikes are fine-grained with diabasic or porphyritic texture and the primary minerals consist of plagioclase, clinopyroxene, and magnetite. Superimposed on the original igneous minerals is a pervasive thermal metamorphism that has altered the primary minerals to a metamorphic assemblage equivalent to the greenschist or prehnite-pumpellyite facies (actinolite, epidote, chlorite, zoisite, prehnite, pumpellyite) (Williams and Malpas, 1972; Duke and Hutchinson, 1974). No deformation has accompanied the low grade thermal metamorphism. The underlying gabbros reveal similar metamorphism but to a much less degree.

The contact between the sheeted dikes and the overlying volcanics is less clearly understood. Williams and Malpas (1972) describe sheeted dikes extending well up into the overlying volcanics. Presumably the volcanic rocks (Table 15) have also been affected by the same thermal metamorphic event and are altered green and red mafic flows and pillow breccias containing minor amounts of keratophyre. Pillow structures in the volcanics indicate their submarine origin. Overlying the pillows is a Lower Ordovician clastic sequence of sandstone, shale, siltstone, and pebble conglomerate (Williams, 1973). Present accounts of the Bay of Islands ophiolite complex do not clearly record any abyssal sediments interlayered with the volcanics nor are any present within the Humber Arm allochthon.

2.3 Petrologic, Tectonic, and Geophysical Considerations

The Bay of Islands ophiolite appears to have preserved a nearly complete record of its development. The internal aspects show that the basal peridotite probably had its origins within the mantle and that the harzburgite-dunite represents a residue derived from a partial fusion process. The structural discontinuity between the transition zone and basal peridotite, along with the rather abrupt change from a uniform and slightly deformed harzburgite-dunite to cumulate layers containing calcic plagioclase, olivine, and clinopyroxene, marks a profound break from deep mantle processes to shallow crystal fractionation of a basaltic liquid. The accessory chromites within the underlying peridotites have a higher Al/Cr ratio than the massive chromites near the transition zone, suggesting that the massive chromites may mark the base of the cumulate zone (Smith, 1958; Irvine and Findlay, 1972). The transition zone and gabbro apparently represent a shallow layered mafic body the formation of which postdates that of the underlying metamorphic peridotite. Chemical and petrologic evidence point to crystal fractionation of a basaltic liquid to form the transition and gabbro zones. There is, however, another unconformity between the dike swarm and gabbro section and this is expressed by apparent deformation of the gabbro prior to the injection of the dike swarm. As shown in earlier chapters, dike swarms are generally interpreted as forming vertical sheets at an axis of spreading (Moores and Vine, 1971) and, if derived from underlying cumulate gabbros, should be normal to this layering at least during the formation of the sheeted dikes. Here we see chilled dikes of diabase extending downward into gabbro whose primary layering has either been rotated or folded and is not perpendicular to the dike swarm trends. Chemical considerations do, however, provide evidence that the gabbros and transition zone rocks were probably derived from the same basic magma as the sheeted dikes and overlying volcanics. The structural discordance between the dike swarm and gabbros demonstrates that a deformational event interrupted the apparent derivation from the same magma. Brecciation within the dikes yields evidence of yet another period of deformation during the formation of the ophiolite complex. All of the internal petrologic features of the ophiolite are seemingly unrelated to the surrounding country rocks and can be considered as having had their origins in the oceanic realm.

Perhaps the most difficult aspect of relating the Bay of Islands ophiolites to an oceanic origin is an explanation for the metamorphic aureole at its base. Field observations by Williams and Smyth (1973) apparently show that the protolith for the metamorphic aureole is supracrustal sedimentary rock; however, to the north, at the Hare Bay complex, the protolith is mafic volcanic rock. If these metamorphic aureoles are the result of a hot peridotite this then requires that the peridotite, which shows a mantle history of partial fusion, retained enough heat during its transport to produce this metamorphic event as it came in contact with the allochthonous Humber Arm slices or that it was detached shortly after its formation. Even though the aureoles are less than 300 m thick, this still requires a temperature of at least 600° C within the peridotite. Alternately, the amphibolites may represent thin slices of mafic oceanic rocks that were metamorphosed in the oceanic lithosphere or along the continental margin prior to the transport of the ophiolite westward. The peculiar mineral assemblages in the lherzolite-ariegite within the peridotite at the metamorphic contact suggest a rather profound deep seated metamorphism and metasomatic exchange. Further detailed mineralogical and petrologic work is required to satisfactorily explain these unusual rocks at the peridotite base.

Considerations of the actual shape of the Bay of Islands ophiolite complex requires other evidence than that provided by geologic mapping. Earlier workers Ingerson (1935, 1937), Cooper (1936), and Smith (1958), regarded the separate ophiolite masses in the Bay of Islands as autochthonous igneous plutons whereas the present interpretation requires that they be rootless allochthonous slabs. A Bouguer gravity profile across North Arm Mountain by Weaver (1967) provides independent evidence as to the shape of these ophiolite masses. Comparing the calculated anomalies with the observed anomaly, it was not possible to arrive at an unequivocal answer. However, the geophysical data suggests that the Bay of Islands ophiolite masses are rootless and are best interpreted as thick horizontal slabs. The conclusion reached from the geologic and geophysical considerations is that the Bay of Islands ophiolite represents a transported slice of oceanic crust obducted over the continental margin during Middle Ordovician.

3. Troodos Ophiolite Complex, Cyprus

3.1 Geologic Situation

In the extreme eastern part of the Mediterranean Sea the Island of Cyprus (9300 km^2) is situated 75 km South the Taurus coastline of Turkey and 105 km West of Latakia on the coast of Syria. On the island, two subparallel, east-west mountain belts are separated by a central plain and these features constitute the main topographic divisions of Cyprus. Folded Mesozoic sedimentary rocks of the Kyrenia range form a gentle arc along the north coast and are considered the southern most portion of the Tauro-Dinaric Alps. To the South, the Troodos range consists of a nearly complete allochthonous ophiolite complex (88 km × 29 km, Fig. 66). The Troodos ophiolite complex has been carefully documented by geologic mapping at the scale of two inches to one mile (Wilson, 1959; Bagnall, 1960; Bear, 1960; Gass, 1960) and a smaller scale geologic map (1 : 250 000) covers the whole island of Cyprus (Bear, 1963 b). This excellent mapping by the Cyprus Geological Survey produced the basic geologic facts that have provided impetus for numerous additional studies, and up to the present time, the Troodos complex is considered the best known ophiolite complex in the world (Gass, 1963; Gass and Masson-Smith, 1963; Moores and Vine, 1971; Peterman et al., 1971; Greenbaum, 1972; Lapierre and Parrot, 1972; Gass and Smewing, 1973; Miyashiro, 1973 a; Magaritz and Taylor, 1974; Menzies and Allen, 1974; Spooner et al., 1974). The Troodos ophiolite complex is thought to be a preserved portion of oceanic crust that formed during the development of Tethyan Seaway in Mesozoic times (Juteau, 1970; Moores and Vine, 1971; Smith, 1971; Gass and Smewing, 1973; Mesorian et al., 1973).

The oldest sedimentary rocks so far identified on Cyprus are deep water umber and radiolarian shales (Perapedhi Formation) resting unconformably on top of the pillow lavas which are considered the highest stratigraphic unit in the Troodos ophiolite complex. These umbers are considered to be Campanian in age (Robertson and Hudson, 1973; Robertson, 1975), and, therefore, place a minimum age on the formation of the volcanics. Abbate et al. (1973) have shown convincingly that many of the other ophiolite sequences of the Tethyan seaway are overlain by Upper Jurassic sequences and radiometric ages presented by Lanphere

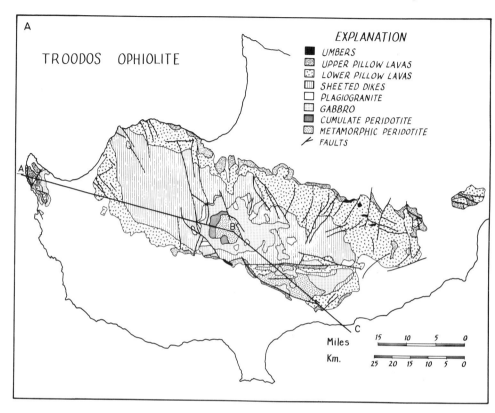

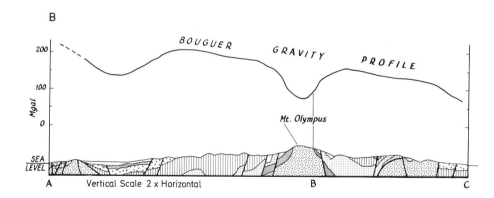

Fig. 66. (A) Geologic map of the Troodos ophiolite, Cyprus. (After Bear, 1963b). (B) Cross section of Troodos ophiolite showing Bouger gravity Profile (Gass and Masson-Smith, 1963)

et al. (1975) on Dinaride Ophiolites give a minimum age of 170 m. y. for their igneous formation. There is no autochthonous basement exposed in Cyprus, but Triassic-Jurassic sediments have been described from mélanges within the Mamonia Formation (Lapierre and Parrot, 1972). Vine et al. (1973) report whole rock K-Ar ages on various ophiolites from 56 to 89 m.y. but these ages must also represent only minimum ages because of the widespread hydrothermal metamorphism. Transgressive Maestrichian carbonate sediments resting on top of the partially eroded Troodos ophiolite complex indicate that its emplacement occurred before late Cretaceous (Mantis, 1970). The emplacement of the Troodos ophiolite complex coincides with a widespread Late Cretaceous orogenic event that has thrust ophiolites (ancient Tethyan oceanic crust) along continental margins within the Alpine-Himalayan belt (Abbate et al., 1973). Thus, it is generally agreed that from Jurassic to Cretaceous times, the Mediterranean (Tethys) formed by sea floor spreading and that most of it was apparently destroyed during or before the Alpine orogeny which involved the collision of the African and Eurasian plates (Smith, 1971). The similarity between the Troodos ophiolite complex and those complexes described from the Apennines (Abbate et al., 1972), Vourinous (Moores, 1969), Othris region (Hynes et al., 1972), the Taurides (Brunn et al., 1970), and the Hatay-Bassit area (Mesorian et al., 1973) is now well established and all of these complexes are considered to represent preserved parts of the old Tethyan oceanic floor.

3.2 Internal Character

The Troodos ophiolite complex can be divided into four units for the purposes of this discussion and from top to bottom they are: (1) Pillow lavas which includes the upper and lower lavas of Bear (1963b); (2) Sheeted diabase including the Basal Group of Wilson (1959); (3) Gabbro and Granophyre section including cumulate ultramafic rocks (Transition Zone); (4) Peridotites (Fig. 67). The peridotites and their serpentinized equivalents are very restricted and are present as small irregular bodies: (1) structural dome at the summit of Mt. Olympus is underlain by a fault bounded core of peridotite; (2) tectonically complex area showing a marked east-west trend within the Limassol forest southeast of Mt. Olympus (Fig. 66). Nowhere is there exposed a contact between the peridotites and older autochthonous rocks. Early workers described the peridotites as intrusive igneous rocks (Wilson, 1959; Bear, 1960) into the surrounding gabbros; however, all of the contacts with these rocks are faulted and marked by sheared serpentinized peridotite. The strong negative gravity anomaly described by Gass and Masson-Smith (1963)

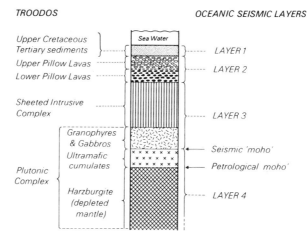

Fig. 67. Correlation between Troodos units and oceanic seismic layers. (After Gass and Smewing, 1973)

at the summit of Mt. Olympus requires a large mass of light material. Sheared serpentinite derived from the peridotite during its emplacement could conceivably produce such a light mass followed by invasion into the overlying volcanics, diabase, and gabbros as a diapir much as a salt dome (Gass and Masson-Smith, 1963; Coleman, 1971 b). The peridotite (Table 16) consists mainly of harzburgite and minor amounts of dunite whose contacts are steeply inclined and show no chilling one against the other. Mineral banding is discontinuous cutting across the boundaries of the major rocks and also dips steeply. The banding is discontinuous usually pinching out within less than 4 m and consists of variable amounts of olivine and clinopyroxene. The banding trends north-south and chromite lenses within the dunites are also elongated in this direction (Wilson, 1959). The dunites (Table 16) consist mainly of olivine (Fo_{92})-chromite and exhibit a xenomorphic granular fabric. The harzburgite (Table 16) contains 80% olivine (Fo_{90-92}), 20% orthopyroxene (En_{90-92} with exsolved diopside), and accessory aluminous Cr spinel and also has a xenomorphic granular fabric (Menzies and Allen, 1974). Small outcrops of plagioclase lherzolite (Table 16) consisting of olivine (65%), clinopyroxene (10%), orthopyroxene (15%), plagioclase (8%), and spinel (2%) are present within the harzburgite near its western boundary with the dunite. The north-south banding is considered to be a metamorphic fabric produced during sub-solidus recrystallization and partial melting (Menzies and Allen, 1974). Serpentinization is pervasive throughout the peridotite mass within Mt. Olympus approaching 80–100% in most of the rocks. On the eastern side of the

Table 16. Average chemical composition and CIPW norms of the various units within the Troodos ophiolite, Cyprus[a]

	1	2	3	4	5	6	7	8	9
SiO_2	40.68	43.73	42.91	51.47	54.62	71.84	53.89	51.21	46.44
Al_2O_3	0.14	0.47	3.64	18.14	15.88	13.27	15.83	15.17	5.98
FeO	8.79	8.19	8.31	6.35	10.09	4.94	11.22	8.49	8.92
MgO	49.38	46.00	40.49	8.56	6.79	1.62	5.71	9.48	32.55
CaO	0.18	0.77	3.47	13.86	7.59	3.45	8.81	10.18	5.07
Na_2O	—	0.01	0.06	0.98	3.40	4.05	2.70	2.50	0.51
K_2O	—	—	0.01	0.12	0.70	0.28	0.59	2.20	0.08
TiO_2	—	0.01	0.02	0.40	0.77	0.49	1.08	0.60	0.29
P_2O_2	—	—	—	—	—	—	—	—	—
MnO	0.17	0.15	0.18	0.12	0.17	0.05	0.17	0.17	0.15
Cr_2O_3	0.41	0.39	0.65	—	—	—	—	—	—
NiO	0.24	0.27	0.25	—	—	—	—	—	—

Normative minerals

	1	2	3	4	5	6	7	8	9
Q	—	—	—	2.9	0.3	33.6	2.9	—	—
C	—	—	—	—	—	0.04	—	—	—
Or	—	—	0.06	0.7	4.1	1.7	3.5	13.0	0.5
Ab	—	0.09	0.5	8.3	28.7	34.3	22.9	18.7	4.3
An	1.3	1.2	9.6	44.7	26.0	17.1	29.3	23.7	13.8
Ne	—	—	—	—	—	—	—	1.2	—
Di	1.9	2.1	6.0	19.5	9.6	—	12.1	21.8	9.0
Hy	8.6	15.5	7.6	23.1	29.7	12.4	27.3	—	19.5
Ol	87.2	80.5	75.2	—	—	—	—	20.3	52.4
Cm	1.0	0.6	1.4	—	—	—	—	—	—
Il	—	0.02	0.04	0.8	1.4	0.9	2.1	1.1	0.6

[a] All analyses normalized to 100% after removal of H_2O, CO_2 and all iron recalculated as FeO.

(1) Dunite, av. 6, (Menzies and Allen, 1974, Table 1). (2) Harzburgite, av. 8, (Menzies and Allen, 1974, Table 1). (3) Plagioclase lherzolite, av. 2, (Menzies and Allen, 1974, Table 1). (4) Gabbros, av. 6, (Coleman and Peterman, 1975, Table 1; Moores and Vine, 1971, Table 4). (5) Diabase dikes, av. 13, (Moores and Vine, 1971, Table 3). (6) Plagiogranite, av. 9, (Coleman and Peterman, 1975, Table 1; Moores and Vine, 1971, Table 4). (7) Lower Pillow Lavas, av. 17, (Moores and Vine, 1971, Table 2). (8) Upper Pillow Lavas, av. 8, (Moores and Vine, 1971, Table 1). (9) Ultrabasic upper pillow lavas, av. 3, (Moores and Vine, 1971, Table 1).

peridotite mass, tectonic movement has produced a large area of sheared serpentinite as part of the diapiric upward movement of the peridotite whereas to the west, the serpentinized peridotites are massive and show only localized areas of shearing. Lizardite, chrysotile, brucite, and magnetite constitute the common serpentinite assemblage with antigorite only rarely found. Oxygen and hydrogen isotopic compositions of the lizardites and chrysotiles are normal with D/H ratios of ($\delta D = -70$ to

−92) but the $\delta^{18}O$ values (+12.6 to +14.1) are extremely high. These results have been interpreted by Magaritz and Taylor (1974) as indicating very low temperature and near surface formation of serpentinite from waters whose origin was meteoric rather than heated ocean waters. This stable isotope data provides further evidence for crustal serpentinization and later diapiric movement of the peridotite during and after its emplacement as part of the Alpine orogenic event.

The pervasive serpentinization and tectonic movement of the peridotites has produced enough disruption so that the relationships with the tectonite peridotite and associated cumulate gabbros and ultramafics cannot be clearly defined in the field (Wilson, 1959; Moores and Vine, 1971). On the western and southern flanks of Mt. Olympus there are extensive outcrops of ultramafic cumulate rocks previously referred to as the peridotite-pyroxenite group (Wilson, 1959) (Fig. 66). These rocks grade downward from the gabbros and contain successively more mafic components, all show banding produced by cumulate processes that develop phase layering of olivine, orthopyroxene, clinopyroxene, olivine, and spinel. Greenbaum (1972) visualizes this sequence as having formed by magmatic differentiation and has worked out the cumulus assemblages from bottom upwards as follows: chromite, olivine + chromite, olivine + clinopyroxene, olivine + clinopyroxene + orthopyroxene + plagioclase, clinopyroxene + orthopyroxene + plagioclase (Fig. 10). These layered rocks develop discontinuous bands and are broken by a series of normal faults formed during the diapiric emplacement of the peridotite core.

Apparently no sharp boundary can be drawn between the metamorphic peridotite and the overlying cumulate sequence (Transitional Zone) and within the sequence, the basal cumulate rocks are gabbro norite, troctolite, wehrlite, pyroxenites, and dunites. Gradationally overlying these rocks are widespread clinopyroxene-bearing gabbros and gabbro norites whose clinopyroxene progressively becomes more uralitized as they approach the plagiogranite zone (Wilson, 1959) (Table 16). Only rarely is banding or layering seen in these gabbros and results from variable concentrations of clinopyroxene and plagioclase. Cumulate textures are developed by the plagioclase (An_{72-92}) laths lying subparallel in the plane of layering associated with stumpy subhedral to euhedral crystals of calcic clinopyroxene and hypersthene. The uralite gabbros grade gradually into plagiogranites (granophyre) with the appearance of quartz and the gradual diminution of pyroxene so that these leucocratic rocks consist primarily of quartz and sodic plagioclase with minor amounts of hornblende and epidote showing a granophyric texture (Wilson, 1959; Bear, 1960; Coleman and Peterman, 1975). The plagiogranites form the roof or screen between the underlying gabbros and the diabase above. The zone is very irregular with no marked chill of the plagiogranites against the gabbros or vice versa but interdigitation of leucogabbros and granophyric quartz-plagioclase rocks often containing inclusions of each other. The plagiogranites are now considered to be the end product of magmatic differentiation of the basaltic melts that first gave rise to the underlying cumulate section of mafic and ultramafic rocks (Peterman et al., 1971).

The widespread occurrence of albite, epidote, chlorite, actinolite, and secondary quartz in the plagiogranite results from a static hydrothermal metamorphism equivalent to the greenschist facies (Gass and Smewing, 1973; Spooner et al., 1974). The thermal metamorphism extends downward into the uralite gabbros where alteration of the pyroxenes to amphibole-chlorite accompanied by saussuritization of the plagioclase marks the lower limits of the thermal metamorphism brought about by hot circulating ocean water at a spreading ridge (Spooner et al., 1974).

The sheeted diabase unit overlying the gabbro granophyre section forms an east-west zone (80 by 23 km) that underlies the main part of the Troodos ophiolite complex. Measurements from the geologic cross-sections (Wilson, 1959; Bear, 1960) show a thickness of 1.2–1.4 km for the sheeted diabase. The central portion of the sheeted diabase consists of a subvertical dike swarm consisting of 100% dikes with no interdike screens (Fig. 68). These dikes extend upward into the overlying basal group where pillow lavas form 3% to 20% interdike screens and finally die out upward into the pillow lavas. The dikes are considered to be feeders for the overlying submarine lavas. At the base of the sheeted diabase, screens of gabbro and plagiogranite are present although most of the dikes have chilled margins against these rocks and do not appear to have been derived directly from the underlying gabbro-granophyre units, nor do they extend downward beyond the upper parts of the uralite gabbro. Wilson (1959) shows gabbroic dikes extending upward into the overlying diabase, however, later workers have interpreted these as coarse-grained dikes injected downward into the earlier formed gabbro-granophyre (Kidd and Cann, 1974). The individual dikes vary in width from 0.3–4.5 m, are subparallel, have a N-S strike, have fine-grained diabasic textures with asymmetric chilled margins. Kidd and Cann (1974) describe the dikes as follows: "It appears that the time lag between dike injections is sufficient to allow the dykes to cool so that each dyke is chilled against those it intrudes, the dykes becoming markedly finer or cryptocrystalline towards their margins (Moores and Vine, 1971). Since many dykes are intruded by subsequent ones, they may be split several times, resulting in the formation of apparently marginless dykes, which are in fact simply the central portions of once

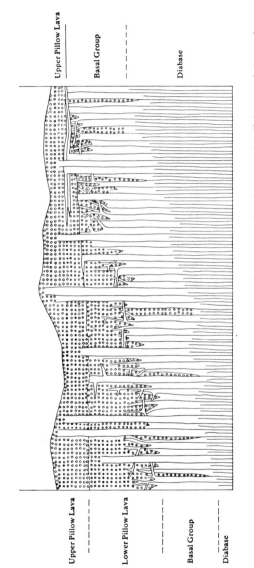

Fig. 68. Diagrammatic sketch showing the relationships, in cross section, between the diabase and the various members of the pillow lava series Troodos, Cyprus. (After Wilson, 1959, Fig. 5)

complete dykes. In some cases a dyke may intrude up the margin of a previous dyke thus creating the situation in which the later dyke is chilled against the chilled margin of a previous dyke." This remarkable conjugation of dikes is now generally interpreted as prima facie evidence of extension and development of new oceanic crust by sea floor spreading at mid-ocean ridges (Moores and Vine, 1971; Kidd and Cann, 1974).

Detailed statistical analysis of asymmetrical chilling margins in the Troodos sheeted diabase complex suggest a spreading center oriented towards the west; however, after rotating 90° counterclockwise to account for the earlier plate movements based on paleomagnetic data of Moores and Vine (1971), the spreading center would have been situated to the north producing a dike swarm whose strike would be E-W (Kidd and Cann, 1974). Even though these dikes retain their original igneous textural characteristics (Table 16), they have undergone nearly complete mineralogical and chemical reconstitution by circulating hot ocean waters (Gass and Smewing, 1973). The diabase now consists mainly of albite, actinolite, chlorite, epidote, and quartz derived from a rock consisting essentially of calcic clinopyroxene, plagioclase (An_{30-60}), and iron oxides. This low grade hydrothermal alteration was accomplished without alteration of the original igneous textures but has considerably altered the original relative amounts of water, the oxidation state of iron, enriched the rocks in O^{18}, and disrupted the initial ratio of Sr^{87}/Sr^{86} (Spooner et al., 1974).

The overlying pillow lavas have been divided into the Upper and Lower units and are separated by an unconformity (Table 16). These lavas completely surround the Troodos dome and are transgressively covered by Maestrichian marine limestones (Fig. 66). The two units are approximately 0.8 km thick and consist mainly of hydrothermally altered pillow lavas. Gass and Smewing (1973) describe these two units as follows: "Upper Pillow Lavas. Generally undersaturated, often olivine-bearing, basalts with more basic varieties (limburgites and picrites) occurring at the top of the sequence. Dykes form less than 10% by volume, absence of silica and celadonite, calcite and analcime common. Lower Pillow Lavas. Mainly oversaturated basalts, often intensely silicified, celadonite common. Dykes, sills and massive flows forming between 30–60% of the outcrop." The Upper Pillow Lavas are considered genetically distinct from the underlying Lower Pillow Lavas and are thought to have formed at a later time (Gass and Smewing, 1973). The thermal metamorphism so pervasive within the sheeted diabase has also affected both the Upper and Lower Pillow Lavas, but the new assemblages indicate lower temperature zeolite facies conditions prevailed. The Upper Pillow Lavas contain natrolite, gmelinite, and phillipsite, whereas laumontite, celadonite, mordenite and chalcedony appear to be restricted to the Lower Pillow Lavas. Thus a progressive downward increase in temperature is recorded by the metamorphic phases contained in the Troodos ophiolite complex (Gass and Smewing, 1973; see Part IV.3).

3.3 Petrologic, Tectonic, and Geophysical Considerations

The Troodos ophiolite complex clearly shows a polygenetic igneous history. The basal peridotites containing essentially olivine and ortho-pyroxene exhibit subsolidus deformation and apparently represent the residue of an early partial melting event with the mantle (Greenbaum, 1972; Menzies and Allen, 1974). There are, however, no traces of the products of partial melting left within the deformed basal peridotite, nor is there evident any feeder channels penetrating the overlying cumulate sequence. The overlying cumulate sequence apparently formed after partial melting and deformation of underlying peridotite. As shown in the section on petrogenesis, the strontium isotopes indicate an extreme difference in age between the cumulate and metamorphic peridotites. Strontium isotope data are not yet available for the metamorphic peridotites from the Troodos ophiolite and so the age relationships between the two ultramafic units is unknown. Available descriptions of the contact between the metamorphic peridotites and the overlying cumulates suggest a complex situation with perhaps the basal dunite cumulates penetrating into the underlying peridotites (see for instance Wilson, 1959, Sheet No. 2). The cumulate section ranges upward from dunites–wehrlites to norites–troctolites–gabbros whose upper parts have differentiated into plagiogranites. Greenbaum (1972, p. 19) visualizes the cumulate sequence as having developed in a magma chamber situated below the axial zone of a slow spreading oceanic ridge and describes it as follows: "At a slight depth beneath the ridge magma exists along its axis as a trough-like pool, fed centrally by a tapering conduit. The enclosing mantle in this zone is residual harzburgite. Crystallization proceeds within the magma chamber and gives rise to a sequence of mafic and ultramafic accumulates. The central zone above the reservoir is one of intrusion of basaltic dykes and extrusion of surface pillowed flows" (Fig. 10). Presumably not all of the liquid is differentiated before injected as dikes or pillow lavas. However, the Upper Pillow Lavas appear to be more mafic than the Lower Pillow Lavas and it has been suggested that the upper Pillow Lavas represent a later extrusion of less differentiated submarine flows developed away from the spreading axis (Gass and Smewing, 1973).

Miyashiro's (1973a) re-evaluation of the chemical analyses of the Lower Pillow Lavas and sheeted dikes had led him to conclude that these rocks have a calc-alkaline trend which best fits their origin within an island arc with thin oceanic crust. Miyashiro's suggestion has brought forth considerable negative reaction (Gass et al., 1975; Hynes, 1975; Moores, 1975) and it would appear that Miyashiro's replies (1975b, c) did not produce a totally convincing argument (Smith, 1975). The lack of sedimentary sequences such as are typical of island arcs (i.e. coarse grained clastic and volcanoclastic rocks typical of arc sequences) in the Cyprus sediments combined with the widespread sheeted dike swarms is geologic evidence incompatible with Miyashiro's suggestion. There is, of course, the real possibility that island arc volcanics may rest on fragments of oceanic crust (Ewart and Bryan, 1973) or could be tectonically imbricated one with the other.

The metamorphic peridotites, the cumulate section, the overlying constructional sheeted dikes and pillows apparently then represent a polygenetic association developed nearly contemporaneously and coalescing to form new oceanic crust at a spreading center. The high heat flow and downward migration of seawater near the spreading center have superimposed a low grade metamorphism on the pillow lavas and sheeted dikes that increases downward to greenschist assemblages (Gass and Smewing, 1973). Isotope geochemistry has fingerprinted the water participating in the low grade metamorphism as sea water, and has led to the conclusion that there must be widespread metamorphism in the upper portions of the oceanic crust near spreading centers (Spooner et al., 1974). The massive copper-bearing sulfide deposits within the Cyprus pillow lavas are considered to have also formed during the low grade hydrothermal metamorphism (Constantinou and Govett, 1973; see Part V.2).

All of the processes described above are considered to have taken place during the development of new Tethyan oceanic crust either at an axial spreading center or within a marginal basin along the Tethyan ocean perimeter. Smith (1971) visualizes the emplacement of oceanic fragments (ophiolites), such as Cyprus, to have taken place as part of the relative movement between Eurasia and Africa during the Late Cretaceous. This relative north-south movement obducted a small fragment of oceanic crust onto the African continent in the vicinity of Cyprus. Bouguer gravity determinations on Cyprus revealed east-west trending positive anomalies between 100–250 mgals situated over the Troodos ophiolite complex (Gass and Masson-Smith, 1963). Current interpretations of the gravity are consistent with the concept that the Cyprus ophiolite complex is a rootless slab of oceanic crust approximately

11 km thick resting on African continental crust (Gass, 1967; Vine et al., 1973). Significantly, a circular negative anomaly of 120 mgal is situated directly over the Mt. Olympus exposure of the ultramafic rocks (Gass and Masson-Smith, 1963). Extensive serpentinization of the ultramafic rocks is the most logical explanation for this peculiar negative anomaly. Tectonic emplacement or movement of ultramafic rocks in the presence of water is conductive to serpentinization (Coleman, 1971 b) and it seems likely that during obduction of the oceanic crust and its later tectonic movements widespread serpentinization occurred. Upward diapiric movement of the serpentinized peridotites must account for the doming of Mt. Olympus rather than igneous processes as originally postulated by Bear (1960). Oxygen and hydrogen isotopic compositions of the serpentine minerals from Mt. Olympus demonstrate that the source of the water that produced the serpentinite was *not* oceanic but probably meteoric water (Magaritz and Taylor, 1974). This is in marked contrast to the stable isotope results determined for the metamorphosed lavas and dikes where it is evident that sea water participated in the alteration. The process of serpentinization appears to have taken place after obduction and during diapiric movement of the Troodos peridotite; probably by invasion of water along the bottom and side contacts of the peridotite after its detachment.

Recent studies by Vine et al. (1973) on the aeromagnetic anomalies of Cyprus combined with measurements of the magnetic properties of the various units from the Troodos ophiolite complex are equivocal. If the ophiolite complex was formed by sea floor spreading, the time represented by distances normal to the strike of the sheeted complex would be approximately 10–15 m. y. of spreading history (assuming 1–2 cm/yr spreading rate). However, the results of this study did not reveal any reversely magnetized material or zones within the Troodos ophiolite complex. These negative results can be explained by the fact that the low grade metamorphism has altered much of the primary magnetite. Also, if the period of spreading was mid to Late Cretaceous, the Earth's magnetic field was of constant and normal polarity at that time (Vine et al., 1973). Thus, these very interesting studies neither confirm or refute by indirect geophysical interpretation the ultimate origin of the Cyprus Ophiolite. In contrast to the magnetic work, a study of seismic velocities of various specific rock units from the complex can be correlated with the seismic velocities measured in the upper parts of the present day oceanic crust (Matthews et al., 1971; Poster, 1973). These continued studies combining geologic, petrologic, geophysical, and geochemical studies on the Troodos ophiolite complex have produced a nearly internally consistent argument that the complex must represent an allochthonous slab of oceanic crust.

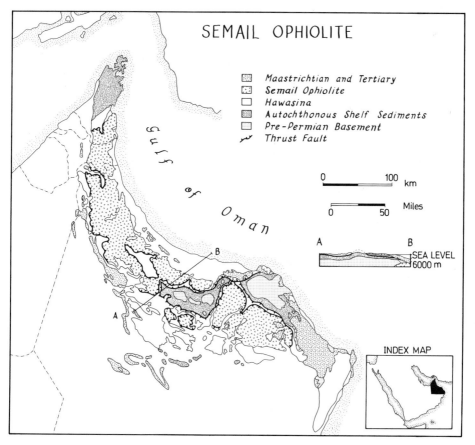

Fig. 69. Geologic map of the Semail ophiolite, Oman. (After Glennie et al., 1974)

4. Semail Ophiolite, Oman

4.1 Geologic Situation

The Semail ophiolite complex is situated within the Sultanate of Oman and the Union of Arab Emirates and forms part of the Oman Mountains extending from the Musandam Peninsula in the north to Ras Al Hadd, the most easterly point of the Arabian Peninsula (Fig. 69). The Semail ophiolites are part of the Middle East alpine mountain chain forming the southernmost part of the "peri-Arabian ophiolite crescent" that can be followed westward from Oman through Neyriz Kermanshah in Iran, along the Turkish-Iran border fold belt, and finally into Hatay (Ricou, 1971). It is now considered that the ophiolites of the peri-Arabian Crescent represent oceanic crust formed in the "Paleo-Tethys" Sea as described by Stocklin (1974): "With the new concepts it is thought that most of the Tethys trough has disappeared, 'consumed' by subduction. What is left of it (the ancient oceanic crust of the Tethys Sea) within the mountain belt are those narrow scars of ophiolites, which are thought to represent former oceanic crust and to define the sutures along which ancient continents or continental fragments have been welded together—i.e. ancient continental margins." The ophiolites within the Alpine mountain chains of the Middle East are now inter-preted as allochthonous slabs that have become imbricated with both shelf carbonates and pelagic cherts-shales during the closing of the Tethys through during Late Cretaceous.

The Semail ophiolite is perhaps the largest ($30,000 \text{ km}^3$) and the best exposed section of ancient oceanic crust within the world. There has been a recent reconnaissance study of the Oman mountains by the Shell Oil Company and from this excellent work numerous papers have been published (Reinhardt, 1969; Glennie et al., 1973, 1974). Allemann and Peters (1972) have presented an interesting account of the Semail ophiolite where it occurs in Union of Arab Emirates. Earlier reports on the Oman mountains were also mainly of reconnaissance nature and provide only sparse information on the Semail ophiolite (Hudson et al., 1954; Hudson and Chatton, 1959; Morton, 1959; Hudson, 1960; Tschopp, 1967; Greenwood and Loney, 1968; Wilson, 1969). Up to now there are no detailed and systematic studies of the ophiolite. Part of this discussion is derived from field work carried out by the author in Oman during 1973–1974.

The geologic situation of the Oman mountains can best be under-stood by dividing the rock units into three groups (Fig. 69): (1) Basement autochthonous rocks represented by Paleozoic and possible Precambrian metamorphic rocks overlain by a thick sequence of shallow water shelf carbonates (Hajar Super-Group, Mid Permian to Cenomanian), which are characteristic for the whole eastern Arabian continental margin. Continental edge and slope deposits consisting of supratidal and open marine reefal facies (Sumeini Group, Permian to Cenomanian). (2) The Semail ophiolite, a thick sequence of ultramafic and mafic rocks thought to represent ancient oceanic lithosphere, and the Hawasina unit of deep water chert, shales, and limestones deposited in the same time span (Permian to Cenomanian) as Group 1 autochthonous sediments. The Hawasina units were deposited in an environment representing a continental rise and ocean basin, and considered to have been situated northeast of the continental margin (Glennie et al., 1974). The Semail ophiolites and Hawasina are allochthonous units tectonically emplaced during the Late Cretaceous above Group 1 sediments with the ophiolite being the highest structural slice. Emplacement of the Hawasina and Semail ophiolite has produced chaotic masses of mélange underlying the ophiolite and containing mountain size exotic blocks of Permian reefal limestone. Metamorphic zones formed at the base of the Semail ophiolite may be related to earlier hot detachment of the ophiolites in the Tethys trough. (3) Thick, shallow water marine limestones of Late Cretaceous to Middle Tertiary age represent a transgressive sequence following the tectonic emplacement of the Semail ophiolites.

All of the units in Groups 1, 2, and 3 have been involved in simple compressional up-folding sometime during the Oligocene and Miocene. This up-folding was followed by late Tertiary normal faulting and recent uplift along the axis of the Oman mountains. Thus the emplacement of the Semail ophiolite coincides with Eo-Alpine orogenesis and marks the closing of the Tethyan sea during the Late Cretaceous.

4.2 Internal Character

The Semail ophiolite can be divided into the following units, modified after Reinhardt (1969), forming a superposed sequence from bottom to the top:
1. Metamorphic zone including garnet amphibolites and greenschist assemblages.
2. Peridotite section includes both tectonized and cumulate ultra-mafic rocks.
3. Transition zone referred to as PG by Reinhardt.
4. Gabbro section includes both cumulate and massive rocks as well as plagiogranites, and encompasses Reinhardt's G and HG units.
5. Diabase dike section.
6. Volcanics (Fig. 70).

The Semail ophiolite does not consist of a continuous sheet or nappe, but is made up of individual plates whose internal structures suggest interplate independence during tectonic emplacement. Internal low angle thrust faults have led to tectonic repetition of the ophiolite sequences and in some instances overturning of the section. Post emplacement vertical faults have offset sequences but some offsets within the gabbro-peridotites do not extend into overlying diabase and gabbro, suggesting pre-emplacement deformation at the Tethys spreading ridge. In a general way there appears to be more deformation and serpentinization of the peridotites on the leading edge of the nappes (west and south) than on their trailing exposed edges (east and north).

The metamorphic zone at the base of the Semail ophiolite is discontinuous and found sporadically throughout the mountains (Allemann

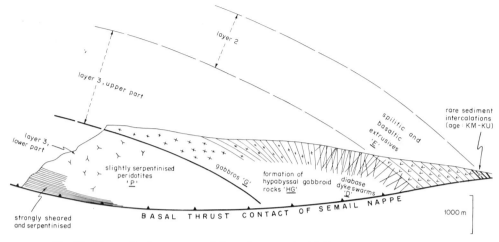

Fig. 70. Generalized structure of Semail nappe (ophiolite) and correlation with ocean-crust layering. (After Glennie et al., 1974, Fig. 6.6.2)

and Peters, 1972; Glennie et al., 1974). Right at the base, the peridotite exhibits a mylonitic texture and where the metamorphic zone is present, the mylonitic peridotite is in contact with garnet amphibolite usually less than 50 m thick. In some areas the amphibolite appears to be present as a tectonic slice and in others the amphibolite appears to grade downward within a short distance to greenschist facies rocks whose protolith appears to be the Hawasina cherts and shales. The presence of these same amphibolites in the mélange units underlying the Semail ophiolite would indicate that their formation was prior to ophiolite emplacement on the Arabian continental platform. Lack of detailed studies of these metamorphic zones prohibits further speculation, but it is worth noting here that the metamorphic aureole described by Williams and Smyth (1973) at the base of the Bay of Islands Ophiolite is petrologically similar. Mélange units are more common than the metamorphics at the base of the Semail ophiolite and consist mainly of Hawasina units immersed in a shaly matrix and it is only rarely that these mélange units contain a serpentinite matrix or broken fragments of the ophiolite.

The peridotites (Table 17) comprise 60% of the outcrop area of the Semail forming a characteristic topography of sharp peaks with a fairly low relief. These rocks have been pervasively serpentinized (60–100%) under static conditions so that the present appearance is of a very dark colored, friable and fractured rock. Harzburgite is the most important original rock type and consists of 60–80% olivine (Fo_{90})

Table 17. Average chemical composition and CIPW norms of the various units within the Semail ophiolite, Oman[a]

	1	2	3	4	5	6
SiO_2	41.95	42.69	46.56	48.34	53.34	57.39
Al_2O_3	0.49	7.82	16.95	23.87	15.44	15.76
FeO	9.86	11.95	5.25	2.82	10.05	8.95
MgO	45.40	28.46	14.26	5.19	6.12	6.47
CaO	0.75	8.02	15.46	17.38	9.43	5.46
Na_2O	0.16	0.40	1.20	2.10	3.91	3.94
K_2O	—	—	—	0.10	0.26	0.59
TiO_2	—	0.10	0.10	0.10	1.28	1.31
MnO	0.11	0.20	—	—	0.10	0.13
Cr_2O_3	0.67	0.13	0.10	0.08	0.04	—
NiO	0.61	0.23	0.12	0.03	0.03	—

Normative minerals

	1	2	3	4	5	6
Q	—	—	—	—	—	5.02
Or	—	—	—	0.6	1.5	3.5
Ab	1.4	1.3	4.8	10.8	33.1	33.3
An	0.6	19.5	40.9	55.4	23.8	23.6
Ne	—	1.1	2.9	3.7	—	—
Di	2.5	16.2	28.6	24.8	19.0	2.9
Hy	5.6	—	—	—	13.6	29.2
Ol	89.0	61.4	22.6	4.3	6.4	—
Cm	1.0	0.2	0.15	0.12	0.06	—
Il	—	0.2	0.19	0.19	2.4	2.5

[a] All analyses normalized to 100% after removal of H_2O, CO_2 and all iron recalculated as FeO.

(1) Peridotites, av. 2, (Glennie et al., 1974, Table 6.8). (2) Cumulate gabbro, av. 3, (Glennie et al., 1974, Table 6.8). (3) Olivine gabbro, av. 3, (Glennie et al., 1974, Table 6.8). (4) Eucrite, av. 3, (Glennie et al., 1974, Table 6.8). (5) Diabase, av. 6, (Glennie et al., 1974, Table 6.8). (6) Pillow lavas, av. 6 (Glennie et al., 1974, Table 6.8).

and 10–25% orthopyroxene (En_{90-91}) with accessory chromite. The harzburgites show a faint layering which is difficult to interpret because of the serpentinization and desert weathering. However, individual hand specimens of fresh unserpentinized material show both tectonic and massive fabrics. Discontinuous lenses of dunite are present within the harzburgite and many contain concentrations of chromite. Ortho-pyroxenite dikes one centimeter to one meter thick commonly cut the harzburgite and dunites. Numerous secondary veins of magnesite are ubiquitous and range in size from centimeters to meters. Up to now there is very little known about the internal structures of the peridotite and it is difficult to establish boundaries between cumulate layering and metamorphic peridotites.

The transition zone marks the first appearance of rocks clearly exhibiting cumulate igneous textures and containing plagioclase. It is characterized by alternating white to dark bands. The dark bands consist of cumulate olivine and clinopyroxene, whereas the light bands consist of anorthosites, troctolites, gabbros, and norites. Below this zone, extending 200 m or more, are cumulate olivine rocks (dunite) some of which contain zones of chromite that also have cumulate structures. The contact between the underlying harzburgites and the cumulates of the transition zone is not clear, but future mapping could provide definite answers. The banded rocks of the transition zone show a more iron-rich olivine (Fo_{85-88}) than found in the underlying harzburgite and the associated plagioclase is very calcic (An_{81-92}) as is the clinopyroxene (Fe 5.8 Mg 50 Ca 44.2) (Reinhardt, 1969). The transition zone exhibits complex relationships with dikes of gabbro, anorthosite, or troctolite cutting the layered sequence which also consists of brecciated parts invaded by leucogabbros. Layering is typically discontinuous and generally pinches out over distances of several meters. The transition zone is further complicated by low angle thrust faults that are localized along the contact between the overlying resistant gabbros and relatively weaker serpentinized peridotites. Reinhardt (1969) reports leucocratic intrusions along these faulted boundaries a process that probably is related to the tectonics at a spreading center.

The transitional zone grades gradually upward into layered gabbros and finally into massive gabbros exhibiting only rare layering (Table 17). The main rock types in this zone are plagioclase (An_{92-65})—rich gabbros containing various proportions of calcic clinopyroxene, olivine (Fo_{72-85}), and orthopyroxene. The layering in these gabbros seems concordant with the underlying transition zone but with fewer cross-cutting dikes. Higher in the section, as the diabase contact is approached, zones of brecciated melagabbro are invaded by leucogabbros. Quartz appears in some gabbros with granophyric textures developing. The sequence from the transition zone to the upper parts of the gabbro demonstrates progressive differentiation of these rocks from melagabbros on up to leucogabbros. Small individual masses of plagiogranite at the top of the gabbro represent the end product of the differentiation and consist essentially of quartz and sodic plagioclase with minor hornblende. The plagiogranites form irregular bodies that crosscut the massive and layered gabbros and where extensive brecciation is present, this same plagiogranite melt has infiltrated the gabbroic breccia.

The contact between the gabbros and overlying sheeted diabase marks a major unconformity in the Semail ophiolite. At the contact, diabase dikes with chilled margins extend downward into the gabbro and crosscut all existing structures. No evidence was found to relate the source of the dikes with the underlying gabbro except by chemical means (Reinhardt, 1969). Large-scale mapping has shown that widespread folding of the gabbros had taken place prior to dike emplacement. If it be assumed that the layering in the gabbros developed as a result of gravity settling, then the plane of layering represents an original horizontal datum plane. Also, if the overlying sheeted dikes represent filling of vertical fractures developed at a spreading ridge, their original configuration must represent a subvertical plane. Assuming the dikes and layered gabbros to be contemporaneous in their initial formation, present day measurements of the dike strike and the plane of the layering in gabbros should produce approximately 90° angles. Instead, lower angles and evidence of folding in the gabbro layering are found. The sheeted dikes of the Semail ophiolite are subparallel, ranging in width from 5 cm to several meters, and exhibiting chilled margins against one another. There are no country rock screens yet reported from these dike swarms and both asymmetric and symmetric chill zones are present, suggesting a mechanism of emplacement similar to that postulated for Cyprus (Moores and Vine, 1971). The sheeted dikes (Table 17) are fine-grained and have an ophitic texture developed by plagioclase An_{40-80}) and clinopyroxene (Fe 9, Mg 45.5, Ca 45.5). Generally the diabase shows an overprint of greenschist facies thermal metamorphism and secondary quartz, albite, actinolite, epidote and chlorite are present as replacement of the original minerals. In some instances a primary brown hornblende is present where advanced differentiation has taken place. Leucocratic dikes having compositions similar to the underlying plagiogranites may have extensive development within the dike swarms and locally predominate over the more mafic dikes.

The volcanics (Table 17) of the Semail ophiolite are the least abundant rocks (3%) and generally form low rounded hills with poor exposures. The dike swarm contact with overlying pillow lavas is marked by increasing screens of lavas between the dikes and anastomosing of the dikes into crosscutting relationships demonstrate that they acted as feeders for the pillow lavas. Nearly all the volcanics exhibit pillow structures and contain abundant interpillow hyoclastic material. Where the attitude of the pillows can be ascertained in relationship to the strike of the sheeted dikes, the angle is approximately 90°. Thus, there seems to be no tectonic break between the pillows and diabase dike swarm. Besides the pillow structures, massive flows, sills, and brecciated lavas are common in the volcanic pile.

The volcanics have all undergone a thermal metamorphism producing zeolite and greenschist assemblages. The preserved textures are dominately intersertal with some porphyritic and variolitic varieties. The feldspar is typically albite associated with chlorite, epidote, quartz and unaltered augite and ilmenomagnetite. Actinolite replaces the pyroxene and the vesicles contain varying proportions of calcite, quartz, and laumontite. The degree of alteration in these rocks is similar to that described in the Cyprus pillow lavas and there appears again to be a downward increase in metamorphic grade into the diabase and gabbro units, but terminating downward somewhere in the gabbro section. Hydrothermal circulation of ocean water near a spreading center is perhaps the best explanation of this alteration (see Part IV. 2). Discovery of widespread copper deposits within the Semail pillow lavas further strengthens its analogy with the Cyprus ophiolite (Bailey and Coleman, 1975; Huston, 1975). Reinhardt (1969) has described mafic and silicic volcanics interbedded with the Hawasina cherts and shales. From his description, combined with my additional field work, it seems clear that these volcanics are not part of the Semail ophiolite sequence developed at a spreading center. The Hawasina volcanics are more mafic in general and appear to have formed as an intrusive event away from the spreading center and may be of different age than the Semail ophiolite. Near the top of the Semail pillow lavas, interlayered ironstones (umbers) are present in many areas, and resemble those described from Cyprus (Robertson, 1975). Sediments in a normal sequence above the Semail pillows are of Cenomanian and Coniacian age and provide a minimum age on the formation of the ophiolite (Glennie et al., 1974).

4.3 Petrologic, Tectonic, and Geophysical Considerations

The petrologic reconstruction of the Semail ophiolite shows that its formation was polygenetic and that these processes probably took place at a spreading center in the Tethys sea and not on the continental margin of the Arabian Peninsula. The basal peridotite has an apparent uniform composition and exhibits evidence of having been deformed under subsolidus conditions. Estimation of temperature and pressure expressed by partition coefficients and pyroxene compositions within the peridotite indicate that temperatures attending peridotite formation were in excess of 1200° C and pressures approached 7 kb (Reinhardt, 1969). The creation of such a large mass of uniform peridotite at these temperatures and pressures requires that this part of the Semail ophiolite formed within the upper mantle. It seems most probable that the Semail peridotite represents a refractory residue formed during a partial melting event in the mantle.

The overlying transitional and gabbro zones are more obviously derived from a fractionating mafic melt. Precipitation of cumulate olivine and chromite signal the beginning of this fractionation. Decreasing $Mg/Mg + Fe$ ratios in the olivines of the gabbros indicates a progressive trend; however, sharp reversals in the cumulate sequence into more mafic layers on top of less mafic layers indicates cyclic events. These cyclic events combined with brecciation, peculiar dike sequences and prediabase faulting, all point to the transient nature of the fractionating magma source. The overlying sheeted dike sequence and pillow lavas, apparently formed at a spreading center, developed after the deformation of the transient underlying cumulate sequence. Thus, there is evidence of three distinct penecontemporaneous events taking place at the spreading center: (1) production of a partial melt residue and its plastic deformation in the mantle; (2) formation of a magma chamber and the development of cumulate sequences by cyclic and random invasions of magma having undergone variable degrees of fractionation; (3) invasion of basaltic and differentiated magmas along vertical tension fractures and outpouring of lavas on the ocean floor. How much time is represented by these three separate events is not known, but the geologic evidence as to their sequence is unmistakable.

The low-grade thermal metamorphism present in the pillow lavas, diabase dikes, and the upper parts of the gabbro is probably related to hot circulating oceanic waters produced by the high heat flow near the spreading centers. This same thermal event may also be responsible for the copper deposits localized in the pillow lavas. Metamorphic zones of amphibolite at the base of the ophiolite may represent the early stages of detachment of a still hot peridotite onto subjacent rocks. All of these events apparently predated the tectonic emplacement of the Semail ophiolite onto the Arabian continental margin. The imbricate thrusting within the ophiolite combined with strong deformation along its leading edge and only mild deformation at its trailing edge, suggests emplacement by gravity sliding. Mélange units under the Semail ophiolite contain mainly Hawasina and Permian sediments in exotic blocks rather than sheared and tectonized ophiolite. Lack of island arc volcanics or trench sediments of Late Cretaceous or earlier age on the Arabian continental margin seems to rule out southward subduction as a geologic process responsible for the emplacement of the Semail ophiolite. Fragmentary gravity and magnetic data on the Semail ophiolite demonstrates that it is a rootless slab with no connection to the mantle underlying the Gulf of Oman.

5. Eastern Papua Ophiolite, New Guinea

5.1 Geologic Situation

The ophiolite in Papua forms a discontinuous belt nearly 400 km long and 40 km wide and is situated on the northeastern side of the Owen Stanley mountain range, the eastern prolongation of New Guinea

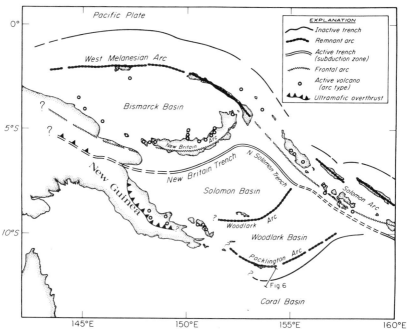

Fig. 71. Generalized tectonic framework of the East Papua—Solomon Islands region. (After Karig, 1972, Fig. 5)

(Fig. 71). The Papuan ophiolite is related tectonically to other Cenozoic ophiolite belts in the Western Pacific, forming a discontinuous line from New Caledonia westward through New Guinea and on into the Celebes, Borneo, and the Philippine Islands (Hutchinson, 1975). The present state of knowledge concerning the interrelated tectonic history of the Western Pacific is still fragmentary and the source of much debate. In this discussion, I will draw heavily on the work of Davies, who has provided an excellent summary and interpretation of the Australian work in this very difficult and inhospitable terrain (Dow and Davies, 1964; Davies, 1968, 1971; Davies and Smith, 1971; England and Davies, 1973). In addition to Davies' work, there are several papers dealing with the structural and geophysical aspects of the Papuan ophiolite (Smith and Green, 1961; Green, 1961; Thompson and Fisher, 1965; Thompson, 1967; Milsom, 1973a, b; Rod, 1974).

The Owen Stanely Range forms a northwest-trending core of metamorphosed Mesozoic sediments and is the basement upon which the Papuan ophiolite rests. These Mesozoic metamorphics consist mostly of graphite-quartz-feldspar-mica schists with marbles, basic schist, metaconglomerates varying in grade from low greenschist in the north to

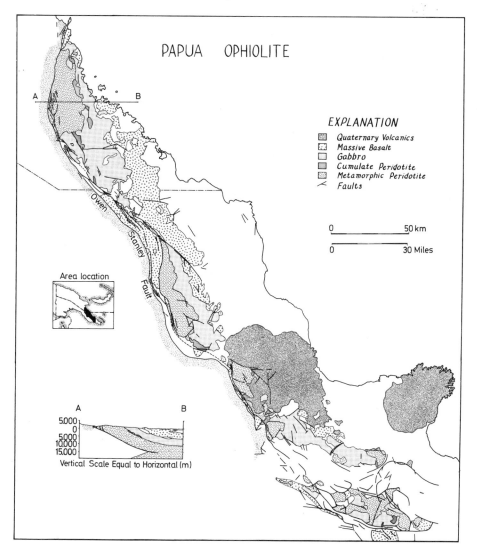

Fig. 72. Geologic map of the Papua—New Guinea ophiolite. (After Davies, 1971)

high greenschist in the south. The Papuan ophiolites and the Owen Stanely metamorphics are separated by a major fault referred to as the Owen Stanely Fault (Fig. 72). Spatially located near the Owen Stanely Fault are sporadic occurrences of lawsonite and glaucophane schists that appear to be related to the emplacement of the ophiolites (Davies and Smith, 1971). Davies (1971) interprets the Owen Stanely Fault as an east-dipping thrust fault extending under the uplifted ophiolites and representing the surface along which the ophiolites were obducted onto continental crust represented by the Owen Stanely Range. Rod (1974) and Milsom (1973a, b) have raised objections to the simple overthrust model. Rod suggests a radical departure to obduction by postulating that the ophiolites have been transported nearly vertically along the Owen Stanely Fault after a west dipping subduction plane underlying the Australian continent had ceased movement during early Tertiary. This requires that the ophiolites be transported vertically from depths of 40 km along reverse faults initially developed during subduction (see for instance Plafker, 1972). The complete lack of high pressure metamorphism within the Papuan ophiolites provides important evidence that these rocks have not experienced deep subduction and later tectonic resurrection.

There are present, however, high grade amphibolite and granulite facies rocks as tectonic slices within the Timeno fault zone east of the main Owen Stanely fault zone; Davies (1971) interprets these rocks as having developed as a result of high temperatures and pressures during thrusting of the Papuan ophiolites (52 m.y. K/Ar age is reported for a hornblende granulite). It is worth noting here that these high grade metamorphics are similar in grade and occurrence to those found at the base of the ophiolites in Newfoundland and Oman. The time of emplacement for the Papuan-New Guinea ophiolite is considered to be upper Eocene or Oligocene as the underthrust and overthrust plates both contain Cretaceous rocks and are covered by transgressive Miocene marine limestones and waterlain pyroclastics. There is a notable lack of interbedded or overlying pelagic sediments or ironstones (umbers) within the upper parts of the pillows. This lack of pelagic sediments may be accounted for by rapid erosion, as it is estimated that during the Miocene considerable vertical movement was initiated and continues up to the present time (Davies, 1971). This Miocene vertical movement has undoubtedly steepened the initial dip of the Owen Stanely thrust fault and also complicated the initial structural relationships with the Papuan ophiolite. If Davies' estimated Eocene ophiolite emplacement is correct, it appears to fit with the reconstruction of plate movement within the western Pacific and he describes this movement as follows: "The northward movement of the Australian plate caused north south compression in the Papua–New Guinea region. The main result was the develoment of the New Britain trench as a sink for the Australian plate; a subsidiary effect may well have been the thrusting of a plate of oceanic mantle and crust (the Papuan Ultramafic Belt), southwards over the Cretaceous geosynclinal sediments of what is now eastern Papua. Westward movement of the Pacific plate undoubtedly has caused east west compression and sinistral torque in the Papua-New Guinea region. The main result has probably been left-lateral strike-slip faulting along the Solomon Islands–New Ireland outer margin; a subsidiary effect may well have been left lateral strike-slip faulting of the Papuan Ultramafic Belt thrust sheet" (see Fig. 71). Rod (1974) has given a similar reconstruction using plate tectonics, but as shown earlier, he resurrects the Papua–New Guinea ophiolite from deep within the subduction zone rather than emplacing it by obduction.

5.2 Internal Character

The Papuan ophiolite is interpreted as a large slab dipping north-east. The distribution of the peridotite, cumulate section, gabbro, and basalt from west to east suggests a grossly layered sequence; however, direct evidence of these successive layered sequences and their eastward dip is not evident in the field. Offsets of the Bouguer gravity highs eastward from the actual outcrop areas provides indirect evidence that Davies' (1971) interpretation of an eastward dipping slab is correct (St. John, 1970; Milsom, 1973 a, b). The internal structure of the ophiolite is poorly known and most of the information is based on geologic transects separated by 6–12 km. It seems likely that structural complications exist such as repetition of section by thrusting or offsets caused by strike-slip or normal faults. For the purposes of this discussion, the ophiolite is divided into the following units from the base upwards: (1) tectonized peridotites; (2) cumulus peridotites; (3) gabbros; (4) basalts. Davies (1971) includes the tectonized peridotites and cumulate peridotites into an ultramafic zone and estimates that 90% of the zone is tectonized peridotite. He also estimates the thickness to vary from 4 to 8 km but cautions that if his estimated dips are too steep, then these thicknesses may be excessive by at least 2 km. I am inclined to believe that a thickness of 2 to 6 km is more reasonable. The degree of deformation is quite variable, but apparently all tectonized peridotites show kink banding of the olivines. Lack of fabric analysis hinders any quantification of the style of deformation, but Davies (1971) reports mylonitic to cataclastic textures. The main rock type is a harzburgite with 60% to 80% olivine ($Fo_{91.6-93.6}$) and 20–40% orthopyroxene ($En_{92.1-93.4}$) and accessory chromite (Table 18). A crude layering is developed by variation in orthopyroxene contents of adjoining bands. Dunite with olivine of the same composition as the harzburgite forms discontinuous layers interlayered with the harzburgite, but locally occurs as crosscutting veins or dikes. Other crosscutting dikes of orthopyroxenite are present and these sometimes enlarge to irregular masses up to 5 m across. Perhaps

Table 18. Average chemical composition and CIPW norms of the various units within the Papuan-New Guinea ophiolite[a]

	1	2	3	4	5	6	7
SiO_2	40.93	43.15	55.52	50.56	51.12	52.12	61.72
Al_2O_3	0.54	0.23	1.13	16.38	18.51	14.00	16.23
FeO	7.82	7.41	9.62	5.26	6.47	11.70	6.87
MgO	49.86	48.46	31.77	12.26	9.15	7.49	3.37
CaO	0.73	0.55	1.62	14.55	12.95	10.42	7.21
Na_2O	—	0.07	0.05	0.73	1.51	2.78	3.52
K_2O	—	0.01	0.04	0.04	0.03	0.07	0.56
TiO_2	—	0.02	0.05	0.11	0.13	1.24	0.41
MnO	0.11	0.10	0.19	0.11	0.13	0.18	0.11

Normative minerals

	1	2	3	4	5	6	7
Q		—	—	—	—	—	15.0
Or		0.06	0.2	0.2	0.2	0.4	3.3
Ab		0.6	0.4	6.2	12.8	23.5	29.8
An		0.3	2.7	41.3	43.6	25.5	26.8
Di		1.9	4.2	24.7	16.7	21.7	7.5
Hy		8.7	85.3	24.7	26.1	23.2	16.8
Ol		88.4	6.9	2.6	0.4	3.2	—
Il		0.04	0.1	0.2	0.3	2.3	0.8

[a] All analyses normalized to 100% after removal of H_2O, CO_2 and all iron recalculated as FeO.

(1) Dunite (Davies, 1971, Table 2). (2) Harzburgite (Davies, 1971, Table 2). (3) Orthopyroxenite, av. 4 (Davies, 1971, Table 2). (4) Gabbro, av. 12, (Davies, 1971, Table 2). (5) High level gabbro (Davies, 1971, Table 2). (6) Basalt, av. 7 (Davies, 1971, Table 2). (7) Tonalite, av. 4, (Davies, 1971, Table 2).

the most important characteristic of the tectonized peridotite is the uniform composition of the olivine and orthopyroxene over great distances (England and Davies, 1973). Such uniformity precludes these rocks representing recrystallized cumulates and strongly favors their origin as refractory residue developed by partial melting in the mantle.

The cumulus peridotite is situated in the upper 100–500 m of the ultramafic zone and forms discontinuous exposures along the entire length of the Papuan ophiolite (Fig. 72). The zone is characterized by finer grain sizes than the underlying tectonized peridotite and contains cumulate olivine, orthopyroxene, clinopyroxene, and chromite with no plagioclase. The olivine ($Fo_{78.3-89.6}$) and orthopyroxene ($En_{81.1-90.5}$) exhibit much wider compositional variations and show a strong iron enrichment. The contact with the overlying gabbros is gradational, whereas the deformation in the underlying peridotite does not penetrate upward into the cumulates and therefore must mark an unconformity within the ophiolites. Davies (1971) reports that serpentinization is widespread and estimates that 20% of the total ultramafic zone is serpentinized. Lizardite and chrysotile are the predominate serpentine species and near fault contacts serpentinization increases rapidly as well as near the tonalite and gabbro contacts. Antigorite is present and appears to be restricted to the Mt. Suckling area and in the ultramafic breccias of the Musa Valley northwest of Mt. Suckling.

Davies (1971) divides the gabbros into three distinct types (Fig. 72): (1) cumulus gabbro which grades into the underlying cumulate peridotite (approximately 3 km); (2) granular gabbro consisting of mainly of a homogeneous mass with minor amounts of cumulus gabbro, streaky gabbro, and gabbro pegmatites (approximately 3 km); and (3) high level gabbro which is massive and characterized by its ophitic textures and zoned plagioclase (approximately 1 km). The cumulate gabbros exhibit rhythmic layering and contain the following cumulate phases: olivine (Fo_{83}), clinopyroxene ($En_{51}Fs_5Wo_{44}$), orthopyroxene (En_{69}) and plagioclase (An_{80-85}). The granular gabbro typically contains 5% olivine (Fo_{80}), 35% clinopyroxene (En_{52}, Fs_5, Wo_{43}), 10% orthopyroxene (En_{83}), 50% plagioclase (An_{80}) and has a fine-grained (1–2 mm) allotriomorphic or hypidomorphic texture. The high level gabbro is situated just below the overlying basalts and grades downward into the granular gabbro. There is no dike swarm situated between the gabbros and basalts as has previously been described in Newfoundland, Cyprus, and Oman. The basal contact of the gabbro is variable. In certain situations gabbro cumulates grade downward into ultramafic cumulates without any structural break. Intrusive contacts are reported by Davies and here the gabbro invades the underlying tectonized peridotite. In a few places the gabbro-tectonized peridotite contact is marked by faults. Postemplacement tonalites invade the tectonized peridotites and overlying gabbros and appear to intrude and enlarge along the basalt gabbro contact.

The basalt zone is a thick (4–6 km) accumulation of massive basalt, pillow basalts, keratophyre and dacitic pyroclastics. In some areas, fine-grained calcareous sediments are interlayered with these volcanics. The lower part of the basalt zone consists of massive diabase and basalt locally invaded by basalt dikes. However, the main part of this zone is characterized by pillow lava flows associated with minor amounts of agglomerate, autobreccias, and hyloclastites. Detailed petrographic descriptions of the basalts are not available, but chemical analyses suggest that they are subalkaline tholeiites (Fig. 72). Davies (1971) reports widespread thermal metamorphism of the basalts with the development of albite, epidote, chlorite, prehnite, and pumpellyite. He attributes this to a possible thermal event produced by hot circulating sea water. Fossils (foraminifera) contained in the interlayered sediments of the basalt zone are considered Late Cretaceous and K/Ar ages on pyroxene from the basalts are also Upper Cretaceous (116 m. y.). Radiometric ages (50–55 m. y.) on the tonalite fix the minimum age of the Papuan ophiolite. However, it is not clear if the tonalite represents a differentiation product of the ophiolite pile or is a later unrelated intrusion that intrude the ophiolite after its emplacement. Important to note here is that no tonalite intrusions invade the Owen Stanely metamorphic rocks, nor do these intrusions crosscut the major fault zones. The chemical characteristics of the tonalites suggest that they may be calc-alkaline and perhaps related to later island arc magmas rather than to the plagiogranites commonly found in ophiolite (Coleman and Peterman, 1975).

5.3 Petrologic, Tectonic, and Geophysical Considerations

The petrologic evolution of the Papuan ophiolitic has a pattern similar to the ophiolite complexes previously described. Here again we see both polygenetic and cogenetic relationships within the complex. The tectonized peridotite has a very uniform mineral composition over large vertical and horizontal distances as well as having been recrystallized at subsolidus temperatures. Its mineral and bulk composition suggest that it formed as a residue of partial melting, rather than as a metamorphosed mantle cumulate (England and Davies, 1973). The low alumina content (1.0% Al_2O_3) of the clinopyroxene from the tectonized peridotite indicates that the partial melting and reequilibration P–T conditions were in the pagioclase peridotite stability field.

In contrast to the tectonized peridotite, the overlying cumulate peridotites, gabbros and pillow basalts apparently have a cogenetic history and formed by magmatic fractionation from the same parent mafic magma. The parent magma may have formed within the mantle by partial melting of a peridotite whose residue conceivably could resemble the underlying tectonized peridotite except for the discrepancies found in the Sr^{86}/Sr^{87} ratios. Davies (1971) has shown that the parent magma must have had a higher Mg/Fe ratio and lower alkali content than continental basalts and more closely resembles typical oceanic basalts. Estimates of parent magma composition by Davies has produced a dilemma, however, as there is no representative of this composition within the lavas overlying the cumulate sequence. The problem here is that amounts of K_2O and TiO_2 in both the parent magma and differentiation products cannot be easily reconciled by simple crystal fractionation of the chosen parent magma. There is apparently too much titanium in the basalts and not enough K_2O.

It would seem that it is not possible to establish the parent magma which itself may be continuously varying in composition as are the relative amounts of the differentiation products.

Davies (1971, p. 39) describes the possible development of the Papuan ophiolite as follows: "The model proposed here is that the ultramafic cumulates and gabbros represent magma reservoirs from which the volcanics were erupted. The reservoirs were fed by a continuing supply of basaltic magma from a mantle source. Reservoirs were between 4 and 10 km below the sea floor when activity ceased, and may have been less than 4 km below the sea floor volcanic activity first started. ... The lack of normal basement makes it difficult to envisage the initial stages in the development of the Papuan complex, but no doubt the process is best seen as one of continuing generation of oceanic crust, such as is believed to be taking place at the midocean ridges today .... In this case the country rock might best be thought of as earlier-formed oceanic crust which moved away from the ridge at a rate of 2–5 cm a year. The gabbro and basalt would then have filled the tensional zone along the axis of the ridge. The noncumulus peridotite would represent mantle material that moved in the solid state, perhaps by syntectonic recrystallisation. This would initially have formed a floor for the gabbroic intrusion, then would have moved with it, as a rigid plate, away from the midocean ridge." The curious lack of dike swarms within the Papuan ophiolite seems to indicate that it represents a different variation of oceanic crustal formation. Karig (1971) has convincingly shown that new oceanic crust forms in marginal basins island arcs and it therefore is possible that the Papuan ophiolite represents a fragment of a marginal basin oceanic crust. The lack of sheeted dikes and the late intrusions of tonalite are geologic indications of a different history for the Papuan ophiolite that conceivably could point to a marginal basin origin.

The allochthonous nature of the Papuan ophiolite appears to be well established using both geologic and geophysical information; however, the emplacement of such a large mass of dense oceanic crust on top of lighter continental crust is a difficult problem in dynamics. Nonetheless, the Papuan ophiolite has been used by numerous authors as a classic illustration of oceanic crust being obducted on to a continental margin (Dewey and Bird, 1971; Coleman, 1971a; Davies, 1971; Davies and Smith, 1971). At the time of emplacement (Paleocene), deep ocean conditions prevailed over much of eastern Papua, thus the site of emplacement was well below sea level. Perhaps as a result of the northward movement of the Australian plate and its convergence with newly formed oceanic crust, a portion of this new oceanic crust was detached and rode over the edge of the Australian continent. It is certainly not clear if the obduction results from gravity sliding or pushing from behind by the oceanic slab. The presence of high pressure–low temperature metamorphic rocks along the Owen Stanley fault zone may suggest a transient subduction of the Australian continent under the oceanic plate. Isostatic adjustment following the burial of light continental crust has elevated the Owen Stanley range to its present position. The original flat lying zone along which the ophiolite was emplaced is now upwarped into a vertical position.

Geophysical studies by Milsom (1973a) shows that the Papuan ophiolites produced positive gravity anomalies similar in size and shape to those discovered in Cyprus for the Troodos ophiolite. Modeling of the Papuan ophiolite as a slab dipping eastward and attached to the present-day oceanic crust and mantle of the Solomon Sea has been successfully carried out by Milsom (1973a, p. 2255) and is summarized as follows: "The hypothesis that the Papuan Ultramafic Belt and similar large ophiolite nappes are slabs of oceanic or frontal-arc crust overthrust together with slices of underlying mantle, onto continental margins is compatible with the observed gravity fields provided that a number of restrictions are accepted. In some respects the allowable structures are so limited that they become difficult to reconcile with the mapped geology and the mechanical conditions for overthrusting, *but since the belt undoubtedly exists, an emplacement mechanism must also.*" (My italics.) The lack of actualistic examples of oceanic crust presently being obducted as part of present day plate movements leaves the emplacement of these allochthonous slabs yet unresolved.

Epilogue

In the process of assembling data for this monograph it became apparent that many aspects of ophiolites have not been fully investigated and to develop rigid conclusions and fixed classifications at this stage would be premature. For this discussion, it would seem most appropriate to indicate some of the major problems connected with the ophiolite concept.

The actual comparison of ophiolite sequences with present-day oceanic lithosphere is incomplete. Up to now there has not been a continuous sampling of the oceanic crust in any one spot. Most of the reconstructions are based on dredge hauls and shallow drilling combined with analogies observed from on-land ophiolites. Until we have sampled complete crustal sections from various localities within the oceans, the analogy between ophiolites and present-day oceanic lithosphere will remain conjectural. Continued deep drilling, combined with dredging in present spreading centers and in structural situations where thick sections of oceanic crust are exposed, will provide lithologic data needed to reinforce the analogy.

The tectonic processes leading to the transfer of oceanic crust into or onto the continental margins remains obscure. Plate tectonic concepts are illustrated by many actualistic situations where such phenomena as subduction, spreading centers, island arcs and other processes can be observed at the present time. However, the transfer of heavy oceanic crust onto lighter continental crust has no present-day counterparts and so the actual mechanics of this process remain obscure. As mentioned earlier, less than 0.001% of the total oceanic crust has been incorporated into the orogens of the continental margins and it is obvious that the tectonic emplacement of these slabs of ocean crust relates to a seemingly unique tectonic situation. Undoubtedly, there will be more than one tectonic situation that will allow the transfer of oceanic crust onto continental margins and continued mapping combined with structural studies should provide fresh ideas.

The petrologic lineage of most ophiolites remains obscure as it is difficult to establish cogenetic relationships between the various parts of the sequence. The metamorphic peridotites at the base of most ophiolites are characteristically depleted in the incompatible elements and must represent the residuum of partial melting in the mantle; however, the significantly high Sr^{87}/Sr^{86} values (0.7015–0.7064) of these rocks indicates that they are much older (> 1.5 b.y.) than the overlying constructional part of the ophiolite. If the parent magma for new oceanic crust results from partial melting in the mantle it is unlikely that the metamorphic peridotites at the base of ophiolites represents the residue of that partial melting. The parent magma for the constructional parts of the ophiolite assemblage appears to be most closely related to the sub-alkaline abyssal oceanic tholeiites, however, as more analytical data is obtained some ophiolites show calc-alkaline affinities and undersatured basaltic magmas have been recognized. These variations in chemistry may be interpreted in various ways and will continue to occupy petrologists seeking the tectonic settings of magma generation. Widespread hydro-thermal alteration of the constructional parts of the ophiolite has modified the original chemistry in these rocks to such a degree that establishment of the parent magma-type may not be possible. Comparison of bulk rock analyses from various ophiolite sequences without regard to degree of alteration can only lead to confusion. Detailed studies of the igneous minerals and their paragenesis within the basaltic magmas that give rise to the ophiolite sequences should prove rewarding. The present petrologic variations recorded in the Phanerozoic ophiolites suggests that the tectonic setting for these rocks may have been diverse. Eventually it may be possible to identify ophiolites from marginal basins, island arcs, or mid-ocean ridges by their petrochemical nature.

Assuming that the petrochemical trends will be significant, it must be emphasized that the associated sediments, tectonic setting, and style of metamorphism must also be considered. The total geologic picture must be developed for each ophiolite before reasonable assumptions can be made regarding its possible petrotectonic history. There still remain many igneous rocks which are part of eugeosynclinal deposits (clastic wedges) and are unrelated to oceanic crust. Continued geologic observations will separate these autochthonous igneous rocks from the allochthonous ophiolites, but careful geologic mapping is required where these rocks are incorporated in mélanges.

The ophiolite concept of Steinmann, modified by plate tectonics and illuminated by modern petrology, remains intact and is essential to our understanding of continental margins, ancient seas, and major suture zones.

References

Abbate,E., Bortolotti,V., Passerini,P.: Introduction to the geology of the Northern Apennines. Sed. Geol. **4**, 207–250, 521–558 (1970)

Abbate,E., Bortolotti,V., Passerini,P.: Studies on mafic and ultramafic rocks. 2-Paleogeographic and tectonic considerations on the ultramafic belts in the Mediterranean area. Boll. Soc. Geol. Italy **91**, 239–282 (1972)

Abbate,E., Bortolotti,V., Passerini,P.: Major structural events related to ophiolites of the Tethys belt. In: Inter. Symp. "Ophiolites in the Earth's Crust". Moscow: Acad. Sci. U.S.S.R. 1973

Allegre,C.J., Montigny,R., Bottinga,Y.: Cortege ophiolitique et cortege oceanique, geochimie comparee et mode degenese. Bull. Soc. Geol. France **15**, 461–477 (1973)

Allemann,F., Peters,T.: The ophiolite-radiolarite belt of the North-Oman Mountains. Eclogae Geol. Helvetiae **65**, 657–697 (1972)

Amann,H., Backer,H., Blissenbach,E.: Metalliferous muds of the marine environment: Offshore Technology Conf. Am. Inst. Mining, Metall. Petrol. Eng. OTC 1759, I 345–I 353 (1973)

Amstutz,G.C.: Spilites and Spilitic Rocks. New York-Heidelberg-Berlin: Springer 1974

Andel,T.H.van, Phillips,J.D., Herzen,R.P.von: Rifting origin for the Vema fracture in the North Atlantic. Earth Planet Sci. Lett. **5**, 296–300 (1969)

Anderson,R.N., Halunen,J.Jr.: Implications of heat flow for metallogenesis in the Bauer Deep. Nature (London) **251** (1974) pp. 473–475

Anonymous: Penrose field conference on ophiolites. Geotimes **17**, 24–25 (1972)

Anonymous: Ophiolites in the earth's crust. Symp. Acad. Sci. U.S.S.R., Geol. Inst. Moscow Abstracts 124 p (1973)

Armstrong,R.L., Dick,H.J.B.: A model for the development of thin overthrust sheets of crystalline rock. Geology **2**, 35–40 (1974)

Arth,J.G., Hanson,G.N.: Quartz diorites derived by partial melting of eclogite or amphibolite at mantle depths. Contrib. Mineral. Petrol. **37**, 161–174 (1972)

Aubouin,J.: Geosynclines. Amsterdam: Elsevier 1965, pp. 1–335

Aumento,F., Loncarevic,B.D., Ross,D.I.: Hudson geotraverse: geology of the MId-Atlantic Ridge at 45° N. Phil. Trans. Roy. Soc. London, Ser. A **268**, 623–650 (1971)

Bachinski,D.J.: Sulfur isotopic composition of ophiolitic cupriferous iron sulfide deposits, Notre Dame Bay, Newfoundland. Econ. Geology **71**, 443–452 (1976)

Bäcker,H.: Rezente hydrothermal—Sedimentäre Lagerstättenbildung. Erzmetall **26**, 544–555 (1973)

Bagnall,P.S.: The geology and mineral resources of the Pano Lefkara-Larnaca area. Mem. Geol. Surv. Cyprus **5**, 1–116 (1960)

Bailey,E.H., Blake,M.C.,Jr.: Major chemical characteristics of Mesozoic coast ophiolite in California. J. Res. U.S. Geol. Survey, **2**, 637–656 (1974)

Bailey,E.H., Blake,M.C.,Jr., Jones,D.L.: On-land Mesozoic oceanic crust in California coast ranges. Geol. Survey Res., 1970, U.S. Geol. Survey Prof. Paper **700**-C, C 70–C 81 (1970)

Bailey,E.H., Coleman,R.G.: Mineral deposits in the Semail ophiolite of northern Oman. Geol. Soc. Am. Abst. 7 (3), 293 (1975)

Bamba,T.: Ophiolite from the Ergani mining district, Southeastern Turkey. Mining Geol. **24**, 297–305 (1974)

Barnes,I., La Marche,V.C.,Jr., Himmelberg,G.: Geochemical evidence of present-day serpentinization. Science **156**, 830–832 (1967)

Barnes,I., O'Neil,J.R.: The relationship between fluids in some fresh Alpine-type Ultramafics and possible modern serpentinization, Western United States. Geol. Soc. Am. Bull. **80**, 1947–1960 (1969)

Barnes,I., Rapp,J.B., O'Neil,J.R.: Metamorphic assemblages and the direction of flow of metamorphic fluids in four instances of serpentinization. Contrib. Mineral Petrol. **35**, 263–276 (1972)

Bear,L.M.: The geology and mineral resources of the Akaki-Lythrodondha area. Cyprus Geol. Surv. Dept. Mem. **3**, 122 (1960)

Bear,L.M.: The mineral resources and mining industry of Cyprus. Cyprus Geol. Surv. Dept. Bull. **1**, 1–208 (1963a)

Bear,L.M.: Geologic Map of Cyprus (1:250,000). Geologic Survey Dept. Cyprus (1963b)

Bearth,P.: Über Eklogite, Glaukophan—Schiefer undmetamorphe Pillowlaven. Schweiz, Miner. Petr. Mitt. **39**, 267–286 (1959)

Bearth,P.: Die Ophiolite der Zone von Zermatt-Saas Fee. Beitr. geol. Karte Schweiz. N.F. **132**, 130 (1967)

Bearth,P.: Gesteins- und Mineral-Paragenesen aus der Ophiolite von Zermatt. Schweiz. Miner. Petr. Mitt. **53**, 299–334 (1973)

Bearth,P.: Zur Gliederung und Metamorphose der Ophiolite der Westalpen. Schweiz. Miner. Petr. Mitt. **54**, 385–397 (1974)

Benson,W.N.: The tectonic conditions accompanying the intrusion of basic and ultrabasic igneous rocks. U.S. Natl. Acad. Sci. Mem. **1**, 1–90 (1926)

Berckhemer,H.: Topographie des „Ivrea-Körpers" abgeleitet aus seismischen und gravimetrischen Daten. Schweiz. Mineral Petrogr. Mitt. **48**, 235–254 (1968).

Berckhemer,H.: Direct evidence for the composition of the lower crust and the Moho. Tectonophysics **8**, 97–105 (1969)

Berger,W.H.: Deep sea sedimentation. In: The Geology of Continental Margins. Burk,C.A., Drake,C.L. (eds.). New York: Springer 1974, pp. 213–241

Bezzi,A., Piccardo,G.B.: Structural features of the ligurian ophiolites: Petrologic evidence for the "Oceanic" floor of the Northern Apennines geosyncline; a contribution to the problem of the Alpine type gabbro-peridotite associations. Mem. Soc. Geol. Ital. **10**, 53–63 (1971)

Bilgrami,S.A.: Mineralogy and petrology of the central part of the Hindubagh igneous complex, Hindubagh mining district, Zhob Valley, West Pakistan. Rec. Geol. Surv. Pakistan **10**, pt. 2-C, 1–28 (1964)

Birch,F.: Speculations on the earth's thermal history. Geol. Soc. Am. Bull. **80**, 133–154 (1965)

Bischoff,J.L.: The Red Sea geothermal deposits: their mineralogy, chemistry and genesis. In: Hot Brines and Recent Heavy Metal Deposits in the Red Sea. Degens,E.T., Ross,D. (eds.). New York: Springer 1969, pp. 368–401

Bischoff,J.L., Dickson,F.W.: Seawater-basalt interaction at 200° C and 500 bars: Implications for origin of sea-floor heavy-metal deposits and regulations of sea-water chemistry. Earth Planetary Sci. Lett. **25**, 387–397 (1975)

Blake, M.C., Jones, D.L., Landis, C.A.: Active continental margins: contrast between California and New Zealand. In: The Geology of Continental Margins. Burk, C.A., Drake, C.L. (eds.). New York: Springer 1974, pp. 853–872

Bonatti, E.: Metallogenesis at oceanic spreading centers. Ann. Rev. Earth Planet. Sci. **3**, 401–431 (1975)

Bonatti, E., Honnorez, J., Ferrara, G.: Equatorial Mid-Atlantic ridge: Petrologic and Sr isotopic evidence for an Alpine-type rock assemblage. Earth Planetary Sci. Lett. **9**, 247–256 (1970)

Bonini, W.E., Loomis, T.P., Robertson, J.D.: Gravity anomalies, ultramafic intrusions, and the tectonics of the region around the Strait of Gibraltar. J. Geophys. Res. **78**, 1372–1382 (1973)

Bottinga, Y., Allegre, C.J.: Thermal aspects of sea-floor spreading and the nature of the oceanic crust. Tectonophysics **18**, 1–17 (1973)

Boudier, F.: Relations Lherzolite—Gabbro-Dunite dans Le massif de Lanzo (Alpes pie montaises): Exemple de fusion pantielle. Ph. D. thesis Nantes 1–106 (1972)

Bowen, N.L.: The origin of ultrabasic and related rocks. Am. J. Sci. **14**, 89–108 (1927)

Bowen, N.L., Tuttle, O.F.: The system $MgO–SiO_2–H_2O$, Geol. Soc. Am. Bull. **60**, 439–460 (1949)

Boyd, F.R.: Garnet peridotites and the system $CaSiO_3–MgSiO_3–Al_2O_3$. Minerlog. Soc. Am. Spec. Paper, **3**, 63–75 (1970)

Boyd, F.R.: A pyroxene geotherm. Geochim. Cosmochim. Acta, **37**, 2533–2546 (1973)

Brongniart, A.: Classification et caractères mineralogiques des roches homogènes et hétérogènes. Paris: F. G. Levrault 1827

Brooks, C., Hart, S.R.: On significance of komatiite, Geology **2**, 107–110 (1974)

Brunn, J.H.: Conditions de gisement des roches basiques en Macedoine accidentale. C. R. Acad. Sci. Paris **210**, 735 (1940)

Brunn, J.H.: Mise en place et differenciation pluto-volcanique du cortège ophiolitique. Rev. Géogr. Phys. Dyn. **3**, 115–132 (1960)

Brunn, J.H.: Les sutures ophiolitiques. Contribution à l'étude des relations entre phénomènes magmatique et orogéniques. Rev. Géogr. Phys. Géol. Dyn. **4**, 89–96, 181–202 (1961)

Brunn, J.H., De Graciansky, P.C., Gutnic, M., Juteau, J., LeFevre, R., Marcoux, J., Monod, O., Poisson, A.: Structure majeurs et corrélations stratigraphiques dans les Taurides occidentales. Bull. Soc. Géol. France **12**, 515–556 (1970)

Buddington, A.F., Hess, H.H.: Layered peridotite laccoliths in the Trout River area, Newfoundland. Am. J. Sci. **33**, 380–388 (1937)

Burtman, V.S., Moldavantsev, Yu.E., Perfiliev, A.S., Schultz, S.S., Jr.: The oceanic crust of the Variscides in the Urals and Tien Shan. In: Developmental Stages of Folded Belts and the Problem of Ophiolites. Part I, Intern. Symp. Ophiolites in the Earth's Crust. Moscow: Acad. Sci. U.S.S.R. 1973, p. 214

Byerly, P.E.: Interpretations of gravity data from the central coast ranges and San Joaquin Valley, California. Geol. Soc. Am. Bull. **77**, 83–94 (1966)

Cady, W.M.: Tectonic setting of the Tertiary volcanic rocks of the Olympic Peninsula, Washington J. Res. U.S. Geol. Survey **3**, 573–582 (1975)

Caillere, S., Kraut, F., Routhier, P.: Etude géologique minéralogique et structurale des gisements et minerais de chrome du massif de Tiébaghi (Nouvelle-Calédonie). Bull. Soc. Geol. France **6**, 169–187 (1956)

Cann, J.R.: Geological processes at mid-ocean ridge crests. Geophys. J. Roy. Astron. Soc. **15**, 331–342 (1968)

Cann, J.R.: Spilites from the Carlsberg Ridge, Indian Ocean, J. Petrol. **10**, 1–19 (1969)

Cann, J.R.: New model for the structure of the ocean crust. Nature (London) **226**, 928–930 (1970)

Cann, J.R.: A model for oceanic crustal structure developed. Geophys. J. Royal Astron. Soc. **39**, 169–187 (1974)

Carmichael, I.S.E., Turner, F.J., Verhoogen, J.: Igneous Petrology. New York: McGraw-Hill, 1974

Challis, G.A.: The origin of New Zealand ultramafic intrusions. J. Petrol. **6**, 322–364 (1965)

Chetelat, E.de: La genèse et l'évolution des gisements de nickel de la Nouvelle-Calédonie. Bull. Soc. Geol. France Ser. 5, **17**, 105–160 (1947)

Chidester, A.H.: Petrology and geochemistry of selected talc-bearing ultramafic rocks and adjacent country rocks in north-central Vermont. U.S. Geol. Surv. Prof. Paper **345**, 1–207 (1962)

Chidester, A.H.: Evolution of the Ultramafic Complexes of Northwestern New England In: Studies of Appalachian Geology—northern and maritime. Zen, E-an, White, W.S., Hadley, J.B., Thompsen, J.B., Jr. (eds.) New York: Interscience 1969, pp. 343–345

Chidester, A.H., Cady, W.M.: Origin and emplacement of Alpine-type ultramafic rocks. Nature Phys. Sci. **240**, 27–31 (1972)

Christensen, N.I.: Fabric, anisotropy, and tectonic history of the Twin Sisters dunite, Washington. Geol. Soc. Am. Bull. **82**, 1681–1694 (1971)

Christensen, N.I., Salisbury, M.H.: Structure and constitution of the lower oceanic crust. Rev. Geophys. Space Physics. **13**, 57–86 (1975)

Church, W.R.: Ophiolite: its definition, origin as oceanic crust, and mode of emplacement in orogenic belts, with special reference to the Appalachians. Dept. Energy, Mines Res. Canada Publ. **42**, 71–85 (1972)

Church, W.R., Stevens, R.K.: Early Paleozoic ophiolite complexes of the New-foundland Appalachians as mantle-oceanic crust sequences. J. Geophys. Res. **76**, 1460–1466 (1971)

Coleman, P.J.: Geology of the Soloman and New Hebrides Islands, as part of the Melanesian re-entrant, Southwest Pacific. Pacific Sci. **24**, 289–314 (1970)

Coleman, R.G.: Jadeite deposits of the Clear Creek area, New Idria district, San Benito County California. J. Petrol. **2**, 209–247 (1961)

Coleman, R.G.: Serpentinites, rodingites and tectonic inclusions in Alpine-type mountain chains. Geol. Soc. Am. Special Paper **73** (1963)

Coleman, R.G.: New Zealand serpentinites and associated metasomatic rocks. New Zealand Geol. Surv. Bull. **76**, 1–102 (1966)

Coleman, R.G.: Low-temperature reaction zones and alpine ultramafic rocks of California, Oregon and Washington. U.S. Geol. Surv. Bull. **1247**, 1–49 (1967)

Coleman, R.G.: Plate tectonic emplacement of upper mantle peridotites along continental edges. J. Geophys. Res. **76**, 1212–1222 (1971a)

Coleman, R.G.: Petrologic and geophysical nature of serpentinites. Geol. Soc. Am. Bull. **82**, 897–918 (1971b)

Coleman, R.G.: The Colebrooke Schist of southwestern Oregon and its relation to the tectonic evolution of the region. U.S. Geol. Surv. Bull. **1339**, 1–61 (1972a)

Coleman, R.G.: Blueschist metamorphism and plate tectonics. 24th Intern. Geol. Congr. Sect. **2**, 19–26 (1972b)

Coleman, R.G.: Ophiolites in the earth's crust: A symposium, field excursions, and cultural exchange in the USSR. Geology **7**, 51–54 (1973)

Coleman, R.G., Blank, H.R., Jr., Hadley, D.G., Fleck, R.J.: A Miocene ophiolite on the Red Sea coastal plain. Trans. Am. Geophy. Union **56**, 1080 (1975a)

Coleman,R.G., Fleck,R.J., Hedge,C.E., Ghent, E.D.: The volcanic rocks of south-west Saudi Arabia and the opening of the Red Sea. U.S. Geol. Surv. Saudi Arabian Project Report **194**, 1975b, 60 p.

Coleman,R.G., Garcia,M., Anglin,C.: The amphibolite of Briggs Creek: A tectonic slice of metamorphosed oceanic crust in south-western Oregon? Geol. Soc. Am. Abst. **8**, 363 (1976)

Coleman,R.G., Irwin,W.P.: Ophiolites and ancient continental margins. In: The Geology of Continental-Margins. Burk,C.A., Drake,C.L. (eds.) New York: Springer 1974, pp. 921–931

Coleman,R.G., Keith,T.E.: A chemical study of serpentinization—Burro Mountain, Californis. J. Petrol. **12**, 311–328 (1971)

Coleman,R.G., Lanphere,M.A.: Distribution and age of high-grade blueschists, associated eclogites and amphibolites from Oregon and California. Geol. Soc. Am. Bull. **82**, 2397–2412 (1971)

Coleman,R.G., Peterman,Z.E.: Oceanic Plagiogranite. J. Geophys. Res. **80**, 1099–1108 (1975)

Constantinou,G., Govett,G.J.S.: Genesis of sulfide deposits, ochre and umber of Cyprus. Inst. Min. Metal. Trans. **81**, B 34–B 46 (1972)

Constantinou,G., Govett,G.J.S.: Geology geochemistry, and genesis of Cyprus sulfide deposits. Econ. Geol. **68**, 843–858 (1973)

Cooke,H.C.: Thetford, Disraeli and Eastern Half of Warwick Map areas, Quebec. Canada Geol. Surv. Mem. **211**, 1–160 (1937)

Cooper,J.R.: Geology of the southern half of they Bay of Islands complex. Newfoundland Dept. Nat. Res., Geol. Sec., Bull. **4**, 1936, 60 p.

Corliss,J.B.: The origin of metal-bearing submarine hydrothermal solutions. J. Geophys. Res. **76**, 8128–8138 (1971)

Cortesogno,L., Gianelli,G., Piccardo,G.B.: Preorogenic metamorphic and tectonic evolution of the ophiolite mafic rocks (Northern Apennine and Tuscany). Boll. Soc. Geol. Italy 1–37 (1975)

Cowan,D.S.: Deformation and metamorphism of the Franciscan subduction zone complex northwest of Pacheco Pass, California, Geol. Soc. Am. Bull. **85**, 1623–1634 (1974)

Cowan,D.S., Mansfield,C.F.: Serpentinite flows on Joaquin Ridge, Southern coast ranges, California. Geol. Soc. Am. Bull. **81**, 2615–2628 (1970)

Dallmeyer,R.D., Williams,H.: $^{40}Ar/^{39}Ar$ release spectra of hornblende from the metamorphic aureole of the Bay of Islands Complex, western Newfoundland: Timing of ohiolite obduction at the ancient continental margin of eastern North America. Geol. Assoc. Canada Ann. Mtg. Prog., Abs., p. 745 (1975)

Dal Piaz,G.V.: Le "granatiti" (rodingiti l.s.) nelle serpentine delle Alpi occidentali italiane. Memorie della Soc. Geol. Italy, **6**, 267–313 (1967)

Dal Piaz,G.V.: Filoni rodingitici e zone di reazione a bassa temperatura al contatto tettonico tra serpentine e rocce incassanti nelle Alpi occidentali italiane. Soc. Italiana di Miner. Petr. **25**, 263–315 (1969)

Dal Piaz,G.V.: Le metamorphisms de haute pression et basse temperature dans l'evolution structurale du bassin ophiolitique alpino-appeninique (2^e partie). Schweiz. Min. Petr. Mitt. **54**, 399–424 (1974)

Dal Piaz,G.V., Hunziker,J.G., Martinotti,G.: La Zona Sesia-Lanzo e l'evoluzione tettonico-metamorfica della Alpi nordoccidentali interne. Mem. Soc. Geol. Italy **11**, 433–460 (1972)

Dana,E.S.: A Textbook of Mineralogy, 4th ed. (revised by W.E.Ford). New York-London: Wiley 1946

Davies,H.L.: Papuan ultramafic belt. 23nd Intern. Geol. Congr. Sect. **1**, 209–220 (1968)

Davies,H.L.: Peridotite-gabbro-basalt complex in eastern Papua: an over-thrust plate of oceanic mantle and crust. Australian Bur. Min. Resur. Bull. **128**, 1971, 48 p.

Davies,H.L., Smith,I.E.: Geology of eastern Papua. Geol. Soc. Am. Bull. **82**, 3299–3312 (1971)

Davis,G.A., Holdaway,M.J., Lipman,P.W., Romey,W.D.: Structure, metamorphism, and plutonism in the south-central Klamath Mountains, California, Geol. Soc. Am. Bull. **76**, 933–966 (1965)

Decandia,F.A., Elter,P.: Riflessioni sul problema della ofioliti nell 'Appennino settentvionale (nota preliminarl). Atti Soc. Tosacana Sc. Natur. (Pisa), Serie A. **76**, 1–9 (1969)

Deffeyes,K.S.: The axial valley: a steady state feature of the terrain. In: Megatectonics of Continents and Oceans. Johnson,H., Smith,B.L. (eds.) New Brunswick: Rutgers Univ. 1970, pp. 194–222

DenTex,E.: Origin of ultramafic rocks, their tectonic setting and history: A contribution to the discussion of the paper "The origin of ultramafic and ultrabasic rocks", by P.J.Wyllie. Tectonophysics **7**, 457–488 (1969)

Dewey,J.F.: Continental margins and ophiolite obduction: Appalachian Caledonian system. In: The Geology of Continental Margins. Burk,C.A., Drake,C.L. (eds.) New York: Springer 1974, pp. 933–950

Dewey,J.F., Bird,J.M.: Mountain belts and the new global tectonics. J. Geophys. Res. **75**, 2625–2647 (1970)

Dewey,J.F., Bird,J.M.: Origin and emplacement of the ophiolite suite: Appalachian ophiolites in Newfoundland: J. Geophys. Res. **76**, 3179–3206 (1971)

Dickey,J.S.,Jr.: Partial fusion products in alpine-type peridotites: Serrannia de Ronda and other examples. Mineral Soc. Am. Spec. Pub. **3**, 33–49 (1970)

Dickey,J.S.,Jr.: A hypothesis of origin for podiform chromite deposits. Geochim. Cosmochim. Acta **39**, 1061–1074 (1975)

Dickey,J.S.,Jr., Yoder,H.S.,Jr.: Partitioning of chromium and aluminium between clinopyroxene and spinel. Carnegie Inst. Wash. Year Book **71**, 384–392 (1972)

Dickinson,W.R.: Plate tectonic models of geosynclines. Earth Planetary Sci. Lett. **10**, 165–174 (1971a)

Dickinson,W.R.: Plate tectonic models for orogeny at continental margins. Nature (London) **232**, 41–42 (1971b)

Dietrich,V., Vuagnat,M., Bertrand,J.: Alpine metamorphism of mafic rocks. Schweiz. Min. Pet. Mitt. **54**, 291–332 (1974)

Dietz,R.S.: Alpine serpentinites as oceanic rind fragments. Geol. Soc. Am. Bull. **74**, 947–952 (1963)

Dow,D.B., Davies,H.L.: Geology of the Bowutu Mountains, New Guinea: Australia Bur. Mineral Resources, Geol. Geophys. Rept. **75**, 1964, 31 p.

Dubertret,L.: Geologie des roches vertes du Nord-Quest de la Syrie et due Hatay. Notes et mem. Moyen-Orient **6**, 2–179 (1955)

Duke,N.A., Hutchinson,R.W.: Geological relationships between massive sulfide bodies and ophiolitic volcanic rocks near York Harbour, Newfoundland. Can. J. Earth Sci. **11**, 53–69 (1974)

Dungan,M.A., Vance,J.A.: Metamorphism of ultramafic rocks from the upper Stillaguamish area, northern Cascades, Washington. Geol. Soc. Am. Ann. Mtg. Abs. **4**, (7) 493 (1972)

Elderfield,H., Gass,I.G., Hammond, A., Bear,L.M.: The origin of ferromanganese sediments associated with the Troodos Massif of Cyprus. Sedimentology **19**, 1–19 (1972)

Elter,P., Trevisan,L.: Olistostromes in the Tectonic Evolution of the Northern Appennines. In: Gravity and Tectonics. DeJong,K.A., Scholten,R. (eds.). New York: Wiley 1973, pp. 175–188

Engin, T., Hirst, D. M.: The alpine chrome ores of the Andizlik-Zimp-Aralik area, Fethyiye. Trans. Instn. Mining Metallurgy (Sec. B: Appl. Earth Sci) **79**, B 16–B 29 (1970)

England, R. N., Davies, H. L.: Mineralogy of ultramafic cumulates and tectonites from eastern Papua. Earth Planet. Sci. Lett. **17**, 416–425 (1973)

Ernst, W. G.: Tectonic contact between the Franciscan melange and the Great Valley sequence—crustal expression of a late Mesozoic Benioff Zone. J. Geophys. Res. **75**, 886–901 (1970)

Ernst, W. G.: Interpretative synthesis of metamorphism in the Alps. Geol. Soc. Bull. **84**, 2053–2078 (1973)

Ernst, W. G.: Metamorphism and ancient continental margins. In: The Geology of Continental Margins. Burk, C. A., Drake, C. L. (eds.). New York: Springer 1974, pp. 907–919

Evans, B. W., Trommsdorff, V.: Regional metamorphism of ultramafic rocks in the Central Alps: Parageneses in the system $CaO-MgO-SiO_2-H_2O$: Schweiz. Mineralog. Petrog. Mitt. **50**, 481–492 (1970)

Ewart, A., Bryan, W. B.: Petrography and geochemistry of the igneous rocks from Eua, Tongan Islands: Geol. Soc. Am. Bull. **80**, 3281–3298 (1972)

Ewart, A., Bryan, W. B.: The petrology and geochemistry of the Tongan Islands: In: The Western Pacific Island Arcs. Marginal Seas, Geochemistry. P. J. Coleman (ed.). Perth, Australia: U. West. Australia 1973, pp. 503–522

Faure, G., Powell, J. L.: Strontium Isotope Geology. New York: Springer 1972

Faust, G. T., Fahey, J. J.: The serpentine-group minerals. U.S. Geol. Surv. Prof. Paper **384**-A, 1–91 (1962)

Flint, D. E., de Albear, J. F., Guild, P. W.: Geology and chromite deposits of the Camaguey district, Camaguey Province, Cuba. U.S. Geol. Surv. Bull. **954**-B, 39–63 (1948)

Fouqué, F., Michel-Levy, A.: Minéralogie micrographique. Mem. Carte Geol. France 153 (1879)

Gale, G. H.: Paleozoic basaltic komatiite and ocean-floor type basalts from northeastern Newfoundland. Earth Planet. Sci. Lett. **18**, 22–28 (1973)

Gansser, A.: Ausseralpine Ophiolit probleme. Eclogae Geol. Helv. **52**, 659–680 (1959)

Gansser, A.: The Insubric line, a major geotectonic problem. Schweiz. Miner. Petr. Mitt. **48**, 123–143 (1968)

Gansser, A.: The ophiolitic melange, a world-wide problem on Tethyan examples. Eclogae Geol. Helv. **67**, 479–507 (1974)

Gass, I. G.: The geology and mineral resources of the Dhali area. Cyprus Geol. Surv. Dept. Mem. **4**, 1960, 116 p.

Gass, I. G.: Is the Troodos Massif of Cyprus a fragment of Mesozoic ocean floor? Nature (London) **220**, 39–42 (1963)

Gass, I. G.: The ultrabasic volcanic assemblages of the Troodos Massif, Cyprus. In: Ultramafic and Related Rocks. Wyllie, P. J. (ed). New York: John Wiley 1967, pp. 121–134

Gass, I. G., Masson-Smith, D.: The geology and gravity anomalies of the Troodos Massif, Cyprus. Roy. Soc. London Philos, Trans., Ser. A **255**, 417–467 (1963)

Gass, I. G., Neary, C. R., Plant, J., Robertson, A. H. F., Simonian, K. O., Smewing, J. D., Spooner, E. T. C., Wilson, R. A. M.: Comments on "The Troodos ophiolitic complex was probably formed in an island arc", by A. Miyashiro and subsequent correspondence by A. Hynes and A. Miyashiro, Earth. Planet, Sci. **25**, 236–238 (1975)

Gass, I. G., Smewing, J. D.: Intrusion, extrusion and metamorphism at constructive margins: evidence from the Troodos massif, Cyprus. Nature (London) **242**, 26–29 (1973)

Glassley, W.: Geochemistry and tectonics of the Crescent Volcanic rocks, Olympic Peninsula, Washington. Geol. Soc. Am. Bull. **85**, 785–794 (1974)

Glennie, K. W., Boeuf, M. G. A., Hughes-Clarke, M. W., Moody-Stuart, M., Pilaar, W. F. H., Reinhardt, B. M.: Late Cretaceous nappes in the Oman Mountains and their geologic evolution. Am. Assoc. Petroleum Geol. Bull. **57**, 5–27 (1973)

Glennie, K. W., Boeuf, M. G. A., Hughes-Clarke, M. W., Moody-Stuart, M., Pilaar, W. F. H., Reinhardt, B. M.: Geology of the Oman Mountains, Part One (Text), Part Two (Tables and Illustrations), Part Three (Enclosures) Kon. Nederlands Geol. Mijb. Gen. Ver. Verh. **31**, 1974, 423 p.

Glikson, A. Y.: Primitive Archaean element distribution patterns: Chemical evidence and geotectonic significance. Earth Planet. Sci. Lett. **12**, 305–320 (1971)

Glikson, A. Y.: Petrology and geochemistry of metamorphosed Archaen ophiolites, Kalgoorlie-Coolgardie, Western Australia. Bur. Min. Res., Geol. Geophys., Canberra **125**, 121–189 (1972)

Goles, G. C.: Trace elements in ultramafic rocks. In: Ultramafic and Related Rocks. Wyllie, P. J. (ed.). New York: Wiley 1967, pp. 222–238

Graciansky, P. de: Existence d'une nappe ophiolotique a l'extrémite occidentale de la chaîne sud. analienne; relations entre les autres unités charriéss et avec des terrains autochtones. C. R. Acad. Sci. **264**, 2876–2879 (1967)

Graciansky, P. de: Le problème des "coloured Melanges" a propos de formations chaotique associées aux ophiolites de Lycie Occidental (Turquie). Rev. Géogr. Phys. et de Geol. Dynam. **15**, 555–566 (1973)

Green, D.: Evolution in meaning of certain geologic terms. Geol. Mag. **108** (2), 177–178 (1971)

Green, D. H.: Ultramafic breccias from the Musa Valley, Eastern Papua. Geol. Mag. **118**, 1–26 (1961)

Green, D. H.: High temperature peridotite intrusions. In: Ultramafic and Related Rocks. Wyllie, P. S. (ed.). New York: Wiley 1967, pp. 212–221

Green, D. H., Nicholis, I. A., Viljoen, M., Viljoen, R.: Experimental demonstration of the existence of peridotitic liquids in earliest Archean magmatism. Geology **3** (1), 11–14 (1975)

Green, D. H., Ringwood, A. E.: The stability fields of aluminous pyroxene peridotite and garnet peridotite and their relevance in upper mantle structure. Earth Planetary Sci. Lett. **3**, 151–160 (1967)

Greenbaum, D.: Magmatic processes at ocean ridges, evidence from the Troodos Massif, Cyprus. Nature Phys. Sci. **238**, 18–21 (1972)

Greenwood, J. E. G. W., Loney, P. E.: Geology and mineral resources of the Trucial Oman Range. Great Britain Inst. Geol. Sci. Overseas Div. 1968, 108 p.

Grow, J. A.: Crustal and upper mantle structure of the central Aleutian arc. Geol. Soc. Am. Bull. **84**, 2169–2192 (1973)

Guillon, J.-H.: Les massifs péridotitiques de Nouvelle-Caledonie-type d'appareil ultrabasique stratiforme de chaine recents. Mémoires Orstom **76**, 1975, 120 p.

Guillon, J.-H., Lawrence, J. L.: The opaque minerals of the ultramafic rocks of New Caledonia. Mineral Deposits **8**, 115–126 (1973)

Guillon, J.-H., Routheir, P.: Les stades d'evolution et de mise en place des massifs ultramafiques de Nouvelle-Caledonie. Bull. du B.R.G.M., Series deuxieme **4**, 5–37 (1971)

Haeri, Y.: Geology of Iran's chromite deposits. In: Symp. Chrome Ore, Ankara. Ankara, Turkey: CENTO 1961, pp. 21–26

Hamilton, W., Mountjoy, W.: Alkali content of alpine ultramafic rocks. Geochim. Cosmochim. Acta **29**, 661–671 (1965)

Harris, P. G., Reay, A., White, I. G.: Chemical composition of the upper mantle. J. Geophys. Res. **72**, 6359–6369 (1967)

Haskin, L. A., Frey, F. A., Schmitt, R. A., Smith, R. H.: Meteoritic, solar and terristrial rare-earth distributions. Phys. Chem. Earth **7**, 167–321 (1966)

Heaton, T. H. E., Sheppard, S. M. F.: Hydrogen and oxygen isotope evidence for the origins of the fluids during the metamorphism of oceanic crust (Troodos complex, Cyprus). NATO Advan. Study Inst. 1–4 (1974)

Heirtzler, J. R., Le Pichon, X.: Crustal structure of the mid-ocean ridges, 3, magnetic anomalies over the mid-Atlantic ridge. J. Geophys. Res. **70**, 4013–4033 (1965)

Helwig, J., Hall, G. A.: Steady state trenches? Geology **2**, 309–316 (1974)

Henckmann, W.: Die Chromerze des Nahen Ostens. Z. Prakt. Geol. **50**, 1–11, 18–24 (1942)

Hess, H. H.: A primary peridotite magma. Am. J. Sci. **35**, 321–344 (1938)

Hess, H. H.: Serpentines, Orogeny, and Epeirogeny. Geol. Soc. Am. Spec. Paper **62**, 391–407 (1955a)

Hess, H. H.: The oceanic crust. J. Marine Res. **14**, 423–439 (1955b)

Hess, H. H.: Mid-oceanic ridges and tectonics of the sea floor. In: Proc. 17th Symp. Colston Res. Society. Univ. Bristol, London: Butterworths 1965, 317–333

Hiessleitner, G.: Serpentin und Chromerz-Geologic der Balkanhalbinsel und eines Teiles von Kleinasien. Aust. J. geol. Bundesanst. Spec. Iss. **1**, part 1, 3–255, part 2, 259–683 (1951–1952)

Himmelberg, G. R., Coleman, R. G.: Chemistry of primary minerals and rocks from the Red Mountain—Del Puerto ultramafic mass, California. U.S. Geol. Surv. Prof. Paper **600**-C, C 18–C 26 (1968)

Holmes, S. W., Stam, J. C., Baley, R.: Target-ophiolites in the Italian Apennines. Northern Miner **60**, 51, 62–63, March 6, 1975

Hopson, C. A., Frano, S. J., Pessagno, E. A., Jr., Mattinson, J. M.: Preliminary report and geologic guide to the Jurassic ophiolite near Point Sal, Southern California coast. Prep. for 71st Ann. Meet. Cordilleran Section. GSA, Field Trip No. 5, 36 p. (1975)

Hotz, P. E.: Nickeliferous laterites in southwestern Oregon and northwestern California. Econ. Geol. **59**, 355–396 (1964)

Hsü, K. J.: Franciscan mélanges as a model for eugeosynclinal sedimentation and underthrusting tectonics. J. Geophys. Res. **76**, 1162–1170 (1971)

Hudson, R. G. S.: The permian and Trias of the Oman Peninsula, Arabia. Geol. Mag. **97**, 299–308 (1960)

Hudson, R. G. S., Chatton, M.: The Musandam limestone (Jurassic to Lower Cretaceous) of Oman, Arabia. Paris, Mus. Nat. d'Histoire Naturelle Notes et Memoires Moyen-Orient **7**, 69–93 (1959)

Hudson, R. G. S., McGugan, A., Morton, D. M.: The structure of the Jebel Hagab area. Trucial Oman. Geol. Soc. London Quart. J. **110**, 121–152 (1954)

Hughes, C. J.: Spilites, keratophyres, and the igneous spectrum. Geol. Mag. **109**, 513–527 (1973)

Hurley, P. M.: Rb^{87}–Sr^{87} relationships in the differentiation of the mantle. In: Ultramafic and Related Rocks. Wyllie, P. J. (ed.). New York: Wiley 1967

Huston, C. C.: Canadian expertise sparks discovery of three copper ore bodies in Oman. Northern Miner **61**, 27, p. 61, Sept. 18, 1975

Hutchison, C. S.: Alpine-type chromite in North Borneo with special reference to Darvel Bay. Am. Mineral. **57**, 835–856 (1972)

Hutchison, C. S.: Ophiolite in southeast Asia. Geol. Soc. Am. Bull. **86**, 797–806 (1975)

Hutchison, C. S., Dhonau, T. J.: Deformation of an alpine ultramafic association in Darvel Bay, East Sabah, Malaysia. Geologie en Mijnbouw, **48**, 481–494 (1969)

Hynes, A.: Comment on "The Troodos Ophiolitic complex was probably formed in an island arc", by A. Miyashiro. Earth Planet. Sci. Lett. **25**, 213–216 (1975)

Hynes, A. J., Nisbet, E. G., Smith, A. G., Welland, M. M. P., Rex, D. C.: Spreading and emplacement ages of some ophiolites in the Othris region (eastern central Greece). Z. Deutsch. Geol. Ges. **123**, 455–468 (1972)

Iishi, K., Saito, M.: Synthesis of antigorite. Am. Mineralogist, **58**, 915–919 (1973)

Ingerson, E.: Layered peridotitic laccoliths in the Trout River area, Newfoundland. Am. J. Sci. **29**, 422–440 (1935)

Ingerson, E.: Layered peridotitic laccoliths in the Trout River Area, Newfoundland: A Reply. Am. J. Sci. **33**, 389–392 (1937)

Irvine, T. N.: Chromian spinel as a petrogenetic indicator, Part 2. Petrologic applications. Can. J. Earth. Sci. **4**, 71–103 (1967)

Irvine, T. N., Findlay, T. C.: Alpine-type peridotite with particular reference to the Bay of Islands complex. Pub. Earth Physics Branch Dept. Energy, Mines Res. Can. **42**, 97–128 (1972)

Irwin, W. P.: Late Mesozoic orogenies in the ultramafic belts of northwestern California and southwestern Oregon. U.S. Geol. Surv. Prof. Paper **501**-C, C 1–C 9 (1964)

Irwin, W. P.: Sequential minimum ages of oceanic crust in accreted tectonic plates of northern California and southern Oregon (abstr.) Geol. Soc. Am. **5**, 63–63 (1973)

Isacks, B., Oliver, J., Sykes, L. R.: Seismology and the new global tectonics. J. Geophys. Res. **73**, 5855–5899 (1968)

Iwao, S.: Albitite and associated jadeite rock from Kotaki, district, Japan; A study in ceramic raw material. Jap. Geol. Surv. Rep. **153**, 1–25 (1953)

Jackson, E. D.: Primary textures and mineral associations in the ultramafic zone of the Stillwater Complex, Montana. U.S. Geol. Surv. Prof. Paper **358**, 1–106 (1961)

Jackson, E. D.: Stratigraphic and lateral variation of chromite composition in the Stillwater Complex. Miner. Soc. Am. Spec. Paper **1**, 46–54 (1963)

Jackson, E. D.: The origin of ultramafic rocks by cumulus processes. Fortschr. Miner. **48**, 128–174 (1971)

Jackson, E. D., Green, H. W. II., Moores, E. M.: The Vourinos ophiolite, Greece: cyclic units of lineated cumulates overlying harzburgite tectonite Geol. Soc. Am. Bull., **86**, 390–398 (1975)

Jackson, E. D., Thayer, T. P.: Some criteria for distinguishing between stratiform, concentric and alpine peridotite-gabbro complexes. 24th Intern. Geol. Congr. Sect. **2**, 289–296 (1972)

Jahns, R. H.: Serpentinites of the Roxbury district, Vermont. In: Ultramafic and Related Rocks. Wyllie, P. J. (ed.). New York: Wiley 1967, pp. 137–160

Jakes, P., Gill, J.: Rare earth elements and the island arc tholeiitic series. Earth Planet. Sci. Lett. **9**, 17–28 (1970)

Johnson, A. E.: Origin of Cyprus pyrite deposits. 24th Intern. Geol. Congr. Sec. **4**, 291–298 (1972)

Jolly, W. T., Smith, R. E.: Degradation and metamorphic differentiation of the Keweenawan tholeiitic lavas of northern Michigan, U.S.A. J. Petrol. **13**, 273–309 (1972)

Juteau, T.: Pétrogenèse des ophiolites des nappes e'Antalya (Taurus lycien oriental Turquie). Science Terre **15**, 265–268 (1970)

Kaaden, G. van der: Chromite-bearing ultramafic and related gabbroic rocks and their relationship to "ophiolitic" extrusive rocks and diabases in Turkey. The Geol. South Africa, Symposium on the Bushveld Igneous Complex, and other layered intrusions, Special Publication **1**, 511–531 (1970)

Karig, D. E.: Origin and development of marginal basins in the western Pacific. J. Geophys. Res. **76**, 2542–2561 (1971)

Karig, D. E.: Remnant arcs. Geol. Soc. Am. Bull. **83**, 1057–1068 (1972)

Kay, M.: North American geosynclines. Geol. Soc. Am. Mem. **48**, 1–143 (1951)

Kay, R. W., Senechal, R. G.: The rare earth chemistry of the Troodos ophiolite complex. J. Geophy. Res. **81**, 964–970 (1976)

Kemp, J. F.: The Mayari iron-ore deposits, Cuba. Am. Inst. Mining Eng. Trans. **51**, 3–30 (1916)

Khain, V. E., Muratov, M. V.: Crustal movements and tectonic structure of Continents. In: The earth's Crust and Upper Mantle. Harte, P. J. (ed.). Am. Geophys. Mono. **13**, 523–538 (1969)

Kidd, R. G. W., Cann, J. R.: Chilling statistics indicate an ocean-floor spreading origin for the Troodos Complex, Cyprus. Earth Planet. Sci. Lett. **24**, 151–155 (1974)

Knipper, A. L.: Osobennosti obrazovaniya antiklinaley s serpentinitovymi yadrami (Sevano—A Kerins Kaya zora Malogo Kavkaza) (characteristics of the development of anticlines with serpentinite cores (Sevan-Akerin zone of the lesser Caucasus)): Moskav. Obshch. Ispytoteley Privody Byull., Otdel Geol. **15**, 46–58 (1965)

Kornprobst, J.: Le massif ultrabasique des Beni Bouchera (Rif. Interne, Maroc). Contr. Mineral. Petrol. **23**, 283–322 (1969)

Kornprobst, J.: Contribution à l'étude petrographique et structurale de la zone enterne du Rif. Ph. D. Thesis Paris 1–376 (1971)

Lamarche, R. V.: Ophiolites of southern Quebec. Pub. Earth Physics Branch Dept. Energy, Mines, Resources. Canada **42**, 65–69 (1972)

Lanphere, M. A.: Strontium isotopic relations in the Canyon Mountain, Oregon and Red Mountain, California ophiolites. Trans. Am. Geophys. Union **54** (11), 1220 (1973)

Lanphere, M. A., Coleman, R. G., Karamata, S., Pamic, J.: Age of amphibolites associated with alpine peridotites in the Dinaride ophiolite zone, Yugoslavia. Earth Planet. Sci. Lett. **26**, 271–276 (1975)

Lapierre, H., Parrot, J.-F.: Identite geologique des regions de Paphos (Chypre) et de Baer-Bassit (Syrie) C. R. Acad. Sci. Paris **274**, 1999–2002 (1972)

Laurent, R.: Occurrences and origin of the ophiolites of southern Quebec, Northern Appalachians. Can. J. Earth Sci. **12**, 443–455 (1975)

Lee, W. H. K., Uyeda, S.: Review of heat flow data. In: Lee, W. H. K. (ed). Terrestrial Heat Flow, Amer. Geophys. Union, Geophys. Monograph Series **8**, 87–190 (1965)

Lemoine, M.: Eugeosynclinical domains of the Alps and the problem of past oceanic areas. 24th Intern. Geol. Congr. Montreal, Sec. 3, 476–485 (1972)

Lockwood, J. P.: Sedimentary and gravity-slide emplacement of serpentinite. Geol. Soc. Am. Bull. **82**, 919–936 (1971)

Lockwood, J. P.: Possible mechanisms for the emplacement of alpine-type serpentinite: Geol. Soc. Am. Mem. **132**, 273–287 (1972)

Loney, R. A., Himmelberg, G. R., Coleman, R. G.: Structure and Petrology of the alpine-type peridotite at Burro Mountain, California, U.S.A. J. Petrol. **12**, 245–309 (1971)

Loomis, T. P.: Contact metamorphism of pelitic rock by the Ronda ultramafic intrusion, southern Spain. Geol. Soc. Am. Bull. **83**, 2449–2474 (1972a)

Loomis, T. P.: Diapiric emplacement of the Ronda high-temperature ultramafic intrusion, southern Spain. Geol. Soc. Am. Bull. **83**, 2475–2496 (1972b)

Loomis, T. P.: Tertiary mantle diapirism, orogeny, and plate tectonics east of the Strait of Gibraltar. Am. J. Sci. **275**, 1–30 (1975)

Lyttle, N. A., Clarke, D. B.: New analyses of Eocene basalt from the Olympic Peninsula, Washington. Geol. Soc. Am. Bull. **86**, 421–427 (1975)

MacDonald, G. A., Katsura, T.: Chemical composition of Hawaiian lavas. J. Petrol. **5**, 82–133 (1964)

MacGregor, I. D.: A study of the contact metamorphic aureole surrounding the Mount Albert ultramafic intrusion. Ph. D. Thesis, Princeton Univ., 1964, 195 p.

MacKenzie, D. B.: High temperature alpine-type peridotite from Venezuela. Geol. Soc. Am. Bull. **71**, 303–318 (1960)

Magaritz, M., Taylor, H. P., Jr.: Oxygen and hydrogen isotope studies of serpentinization in the Troodos ophiolite complex, Cyprus. Earth Planet Sci. Lett. **23**, 8–14 (1974)

Magaritz, M., Taylor, H. P., Jr.: Oxygen, hydrogen and carbon isotope studies of the Franciscan Formation, Coast Ranges, Calif. Geochim. Cosmochim. Acta **40**, 215–234 (1976)

Malpas, J., Stevens, R. K., Strong, D. F.: Amphibolite associated with Newfoundland ophiolite its classification and tectonic significance. Geology **1**, 45–47 (1973)

Manson, V.: Geochemistry of basaltic rocks: Major elements. In: Hess, H., Poldervaart, A. (eds.). Basalts—The Poldervaart Treatise on Rocks of Basaltic Composition. New York-London-Sydney: Interscience Vol. I, pp. 2–270

Mantis, M.: Upper Cretaceous—Tertiary foraminiferal zones in Cyprus. Cyprus Res. Center **111**, 227–241 (1970)

Matthews, D. H., Lort, J., Vertue, T., Poster, C. K., Gass, I. G.: Seismic velocities at the Cyprus outcrop. Nature Phys. Sci. **231**, 200–201 (1971)

Mattinson, J. M.: Early Paleozoic ophiolite complexes of Newfoundland Isotopic ages of zircons. Geology **3**, 181–183 (1975)

Maxwell, J. C.: The Mediterranean, ophiolites and continental drift. In: Megatectonics of continents and oceans. Johnson, H., Smith, B. L. (eds.). New Brunswick, New Jersey: Rutgers Univ. 1970, pp. 167–193

Maxwell, J. C.: Ophiolites—old oceanic crust or internal diapirs? In: Symp. Ophiolites in the Earth's Crust. Moscow: Acad. Sci. USSR 1973, pp. 71–73

Maxwell, J. C.: Early western margin of the United States. In: The geology of Continental Margins. Burk, C. A., Drake, C. L. (eds.) New York: Springer 1974a, pp. 831–852

Maxwell, J. C.: Anatomy of an orogen. Geol. Soc. Am. Bull. **85**, 1195–1204 (1974b)

McDougall, I.: Geochemistry and origin of the Columbia River Basalts of Oregon and Washington. Geol. Soc. Am. Bull. **87**, 777–792 (1976)

McMurtry, G. M., Burnett, W. C.: Hydrothermal metallogenesis in the Bauer Deep of the southeastern Pacific. Nature (London) **254**, 42–43 (1975)

Medaris, L. G., Jr.: High pressure peridotites in southwestern Oregon. Geol. Soc. Am. Bull. **83**, 41–58 (1972)

Melson, W. G., van Andel, T. H.: Metamorphism in the mid-Atlantic Ridge, 22°N latitude. Marine Geol. **4**, 165–186 (1966)

Menzies, M., Allen, C.: Plagioclase lherzolite-residual mantle relationships within two eastern Mediterranean ophiolites. Contr. Mineral. Petrol. **45**, 197–213 (1974)

Mercier, J., Vergely, P.: Les Mélanges ophiolithiques de Macédoine (Grece): décrochements d'âge anté-crétace superieur). Deut. Geol. Gesell. Zeitschr., **123**, 469–489 (1972)

Merritt, P. C.: California asbestos goes to market: Mining Eng. **14**, 57–60 (1962)

Mesorian, H., Juteau, T., Lapierre, H., Nicolas, A., Parrot, J.-F., Ricou, L.-E., Rocci, G., Rollet, M.: Idées actuelles sur la constitution, l'origine et l'évolution des assemblages ophiolitiques mésogeens. Bull. Soc. Géol. France **15**, 478–493 (1973)

Milovanovic, B., Karamata, S.: Über den diaprismus serpentinischer massen. 21st. Intern. Geol. Congr. Copenhagen **18**, 409–417 (1960)

Milsom,J.: Papuan ultramafic belt: gravity anomalies and the emplacement of ophiolites. Geol. Soc. Am. Bull. **84**, 2243–2258 (1973a)

Milsom,J.S.: Gravity field of the Papuan Peninsula. Geologie en Mijnbouw **52**, 13–20 (1973b)

Miyashiro,A.: Evolution of metamorphic belts. J. Petrol. **2**, 277–311 (1961)

Miyashiro,A.: Pressure and temperature conditions and tectonic significance of regional and ocean floor metamorphism. Tectonophysics **13**, 141–159 (1972)

Miyashiro,A.: The Troodos ophiolitic complex was probably formed in an island arc. Earth Planet. Sci. Lett. **19**, 218–224 (1973a)

Miyashiro,A.: Metamorphism and metamorphic belts. London: Allen and Unwin 1973b, pp. 1–479

Miyashiro,A.: Classification, characteristics, and origin of ophiolites. J. Geol. **83**, 249–281 (1975a)

Miyashiro,A.: Origin of the Troodos and other ophiolites: a reply to Hynes. Earth Planet. Sci. Lett.**25**,217–222 (1975b)

Miyashiro,A.: Origin of the Troodos and other ophiolites: a reply to Moores. Earth Planet. Sci. Lett. **25**, 227–235 (1975c)

Miyashiro,A., Shido,F., Ewing,M.: Diversity and origin of abyssal tholeiite from the Mid-Atlantic Ridge near 24° and 30° North Latitude. Contr. Mineral. Petrol. **23**, 38–52 (1969)

Miyashiro,A., Shido,F., Ewing,M.: Petrologic models for the Mid-Atlantic Ridge. Deep-Sea Res. **17**, 109–123 (1970)

Miyashiro,A., Shido,F., Ewing,M.: Metamorphism in the Mid-Atlantic Ridge near 24° and 30°N. Roy. Soc. London Phiolos. Trans. A **268**, 589–603 (1971)

Montigny,R., Bougault,H., Bottinga,Y., Allegre,C.J.: Trace element geochemistry and genesis of the Pindos ophiolite suite. Geochim. Cosmochim. Acta **37**, 2135–2147 (1973)

Montigny,R., Javoy,M., Allegre,C.J.: Sr^{87}/Sr^{86} and O^{18}/O^{16} ratios in the Pindos ophiolitic complex, Greece (abs.) Geol. Soc. Am. **2**, 627–628 (1970)

Moore,J.G.: Mechanism of formation of pillow lava. Am. Sci. **63**, 269–277 (1975)

Moores,E.M.: Petrology and structure of the Vourinous ophiolitic complex, northern Greece. Geol. Soc. Am. Spec. Paper **118**, 1–74 (1969)

Moores,E.M.: Discussion of "Origin of Troodos and other ophiolites: A reply to Hynes" by Akiho Miyashiro. Earth Planet. Sci. Lett. **25**, 223–226 (1975)

Moores,E.M., Jackson,E.D.: Ophiolites and oceanic crust. Nature (London) **250**, 136–138 (1974)

Moores,E.M., Vine,F.J.: Troodos Massif, Cyprus and other ophiolites as oceanic crust: evaluation and implications. Roy. Soc. London Philos. Trans. A **268**, 443–466 (1971)

Morton,D.C.: The geology of Oman. 5th World Petrology Cong. Proc., New York Sect. 1, 277–294 (1959)

Mumpton,F.A., Thompson,C.S.: Mineralogy and origin of the Coalinga asbestos deposit. Clays and Clay Miner. **23**, 131—143 (1975)

Naldrett,A.J.: Archean ultramafic rocks. In: Irving,E. (ed.): The Ancient oceanic Lithosphere. Dept. of Energy, Mines Res. Pub. Earth Physics Br. **42**, 141–151 (1972)

Naldrett,A.J.: Nickel sulphide deposits—their classification and genesis, with special emphasis on deposits of volcanic association. Trans. Can. Inst. Min. Met. **76**, 183–201 (1973)

Naldrett,A.J., Mason,G.D.: Contrasting Archean ultramafic igneous bodies in Dundonald and Clerque Townships, Ontario, Canada. J. Earth Sci. **5**, 111–143 (1968)

Nesbitt, R. W.: Skeletal crystal forms in the ultramafic rocks of the Vilgarn Block, Western Australia: evidence for an Archean ultramafic liquid. Spec. Publ. Geol. Soc. Australia **3**, 331 (1971)

Nickel, E. H.: The occurrence of native nickel-iron in the serpentine rock of the eastern townships of Quebec Province: Can. Mineral. **6**, 307 (1959)

Nicolas, A., Bouchez, J. L., Boudier, F.: Interpretation cinématique des deformations plastiques dans le massif de lherzolite de Lanzo. Tectonophysics **14**, 143–171 (1972)

Nicolas, A., Bouchez, J. L., Boudier, F., Mercier, J. C.: Textures, structures and fabrics due to solid state flow in some European lherzolites. Tectonophysics **12**, 55–86 (1971)

Nicolas, A., Jackson, E. D.: Repartition en deux provinces des péridotites des chaines alpines longeant la Méditerranee: implications géotectoniques. Bull. Suis. Min. Petr. **52**, 479–495 (1972)

Nishimura, M., Yagi, K., Yamamoto, M.: Nickel content of olivines. Proc. Japan. Acad. **44**, 686–691 (1968)

Nochi, M., Komatsu, M.: Ultrabasic rocks in the Hidaka Metamorphic Belt, Hokkaido, Japan II. Earth Science **21**, 11–26 (1967) (in Japanese with English abstract)

O'Connor, J. T.: A classification of quartz-rich igneous rocks based on feldspar ratios. U.S. Geol. Surv. Prof. Paper **525** B, B 79–B 84 (1965)

O'Hara, M. J.: Are ocean floor basalts primary magma? Nature (London) **220**, 683–686 (1968)

Oliver, J. L., Sykes, L., Isacks, B.: Seismology and the new global tectonics. Tectonophysics **7**, 527–541 (1969)

Oxburgh, E. R.: Flake tectonics and continental collision. Nature (London) **239**, 202–204 (1972)

Oxburgh, E. R.: The plain man's guide to plate tectonics. Proc. Geol. Assoc. **85**, 299–357 (1974)

Oxburgh, E. R., Turcotte, D. L.: Thermal gradients and regional metamorphism in overthrust terrains with special reference to the eastern Alps. Schweiz. Miner. Petr. Mitt. **54**, 641–716 (1974)

Page, N. J.: Serpentinization at Burro Mountain, California. Contr. Mineral. Petrol. **14**, 321–342 (1967)

Pamic, J., Scavnicar, S., Medjimorec, S.: Mineral assemblages of amphibolites associated with alpine-type ultramafics in the Dinaride ophiolite zone (Yugoslavia). J. Petrol. **14**, 133–157 (1973)

Parrot, J.-F.: Petrologie de la coupe du Djebel Moussa, massif basique—ultrabasique du Kizil Dag (Hatay, Turquie). Science de la Terre, Nancy **18**, 143–172 (1973)

Passerini, P., Sguazzoni, G.: Ricerche sulle ofioliti delle catene Alpine. 2—Giacitura delle ofioliti nella zona a sud-ovest di Konya (Anatolia meridionale). Boll. Soc. Geol. Ital. **85** (2) 509–523 (1966)

Pearce, J. A.: Basalt geochemistry used to investigate past tectonic environments on Cyprus. Tectonophysics **25**, 41–67 (1975)

Pearce, J. A., Cann, J. R.: Ophiolite origin investigated by discriminant analysis using Ti, Zr and Y. Earth Planet Sci. Lett. **12**, 339–349 (1971)

Pearce, J. A., Cann, J. R.: Tectonic setting of basic volcanic rocks determined using trace element analyses. Earth Planet Sci. Lett. **19**, 290–300 (1973)

Peive, A. V.: Abyssal fracturing in geosyncline regions. Bull. USSR Acad. Sci., Geol. Ser. 1945 **5**, 23–46 (1945)

Peive, A. V.: Oceanic crust of the geologic past. Symp. ophiolites in the earth's crust. Acad. Sci., S.S.S.R. Geol. Inst. Moscow, Pt. I 3–45 (1973)

Peselnick,L., Nicolas,A., Stevenson,P.R.: Velocity anisotropy in a mantle peridotite from the Ivrea Zone: application to upper mantle anisotropy. J. Geophys.
Res. 79, 1175–1182 (1974)

Peterman,Z.E., Coleman,R.G., Hildreth,R.A.: Sr^{87}/Sr^{86} in mafic rocks of the
Troodos Massif, Cyprus. U.S. Geol. Surv. Proof. Paper 750-D, D 157–D 161
(1971)

Peterson,J.J., Fox,P.J., Schreiber,E.: Newfoundland ophiolites and the geology
of the oceanic layer. Nature (London) 247, 194–196 (1974)

Plafker,G.: Alaskan earthquake of 1964 and Chilean earthquake of 1960:
implications for arc tectonics. J. Geophys. Res. 77, 901–915 (1972)

Poldervaart,A.: Chemistry of the earth's crust. Geol. Soc. Am. Special Paper 62,
119–145 (1955)

Poster,C.K.: Ultrasonic velocities in rocks from the Troodos Massif, Cyprus.
Nature Phys. Sci. 243, 2–3 (1973)

Prinz,M.: Geochemistry of basaltic rocks: Trace Elements. In: Hess,H., Poldervaart,A. (eds.): Basalts—The Poldervaart Treatise on Rocks of Basaltic
Composition. New York-London-Sydney: Interscience 1967, Vol. 1, pp. 271–324

Ragan,D.M.: Emplacement of the Twin Sisters dunite, Washington. Am. J. Sci.
261, 549–565 (1963)

Raleigh,C.B., Paterson,M.S.: Experimental deformation of serpentinite and its
tectonic implications. J. Geophys. Res. 70, 3965–3985 (1965)

Ramp,L.: Chromite in southwestern Oregon. Oregon Dept. Geol. Mineral Ind.
Bull. 52, 1–169 (1961)

Reed,J.J.: Chemical and model composition of dunite from Dun Mountain,
Nelson. New Zealand J. Geol. Geophys. 2, 916–919 (1959)

Reim,K.M., Munro,R.G.: Coalinga—Newcomer to the asbestos industry. Min.
Enging. N. Y. 14, 60–62 (1962)

Reinhardt,B.M.: On the genesis and emplacement of ophiolites in the Oman
Mountains geosyncline. Schweizer. Mineralog. Petrog. Mitt. 49, 1–30 (1969)

Reverdatto,V.V., Sobolev,V.C., Sobolev,N.V., Ushakova,Ye.N., Khlestov,V.V.:
Distribution of regional metamorphism facies in U.S.S.R. Intern. Geol. Rev.
8, 1335–1346 (1967)

Reynolds,C.D., Havryluk,I., Saleh Bastaman, Soepomo Atmowidjojo: The exploration of the nickel laterite deposits in Irian Barat, Indonesia. Geol. Soc.
Malaysia Bull. 6, 309–323 (1973)

Ricou,L.E.: Le croissant ophiolitique per-Arabe, une ceinture de nappes mises en
place au Cretace Superieur. Revue de Geographic Phys. Geol. Dyn. 13,
327–349 (1971)

Riordon,P.H.: Preliminary report on the Thetford Mines—Black Lake areas,
Frontenac, Megantic and Wolfe counties. Quebec Depart. Mines Preliminary
Report 295, 1–23 (1954)

Riordon,P.H., Laliberte,R.: Asbestos deposits of southern Quebec. 24th Intern.
Congress Canada. Guide Book Excursion B-08, 1–21 (1972)

Robertson,A.H.F.: Cyprus umbers: basalt-sediment relationships on a Mesozoic
ocean ridge. J. Geol. Soc. London 131, 511–531 (1975)

Robertson,A.H.F., Hudson,J.D.: Cyprus Umbers; chemical precipitates on a
Tethyan Ocean ridge. Earth Planet Sci. Lett. 18, 93–101 (1973)

Rocci,G.: Les ophiolites alpines de méditerranée orientale: uniformité du plutonis-
me, diversité du volcanisme. Intern. Symp. Ophiolites USSR, 1–14 (1973)

Rod,E.: Geology of Eastern Papua: discussion. Geol. Soc. Am. Bull. 85, 653–658
(1974)

Rodgers,K.A.: Ultramafic and related rocks from southern New Caledonia.
Ph. D. Thesis, Auckland, 1972

Rodgers, K.A.: A comparison of the geology of the Papuan and New Caledonian ultramafic belts. J. Geol. **83**, 47–60 (1975)

Rodgers, J., Neale, E.R.W.: Possible "Taconic" Klippen in western Newfoundland. Am. J. Sci. **261**, 713–730 (1963)

Roe, G.D.: Rubidium-strontium analyses of ultramafic rocks and the origin of peridotites. 12th Ann. Rep. M.I.T., 159–190 (1964)

Roeder, D.H.: Subduction and orogeny. J. Geophys. Res. **78**, 5005–5024 (1973)

Roever, W.P. von, de: Sind die alpinotypen Peridotitmassen vielleicht tektonisch verfrachtete Bruchstücke der Peridotitschale? Geol. Rundschau **46**, 137–146 (1957)

Rossman, D.L., Fernandez, N.S., Fontanos, C.A., Zepeda, Z.C.: Chromite deposits on Insular chromite Reservation Number One, Zambales, Philippines. Philippine Bur. Mines Spec. Proj. Pub. **19**, 1–12 (1959)

Routhier, P.: Les gisements de fer de la Nouvelle-Caledonie. 19th Intern. Geol. Cong. Symposium sur les gisements de fer du monde **11**, 567–587 (1952)

Routhier, P.: Les gisements métallifères; géologie et principles de recherche. Paris, Masson et Cie, 1963, Vol. I, 867 p.

Searle, D.L.: Mode of occurrence of the cupriferous pyrite deposits of Cyprus. Inst. Mining Metallurgy Trans. **81**, B 189–B 197 (1972)

Searle, D.L., Vokes, F.M.: Layered ultrabasic lavas from Cyprus. Geol. Mag. **106**, 515–530 (1969)

Shiraki, K.: Metamorphic basement rocks of Yap Islands, western Pacific: possible oceanic crust beneath an island arc. Earth Planet Sci. Lett. **13**, 167–174 (1971)

Shor, G.G., Jr., Raitt, R.W.: Explosion seismic refraction studies of the crust and upper mantle in the Pacific and Indian Oceans. In: The Earth's Crust and Upper Mantle. A.G.U. Geophys. Monograph **13**, 225–230 (1969)

Sigvaldason, G.E.: Epidote and related minerals in two deep geothermal drill holes. Reykjavik and Hveragerdi, Iceland. U.S. Geol. Surv. Prof. Paper **450**-E, E 77–E 84 (1962)

Sillitoe, R.H.: Formation of certain massive sulphide deposits at sites of sea-floor spreading. Inst. Mining Metallurgy Trans. **81**, 141 B–148 B (1972)

Sillitoe, R.H.: Environments of formation of volcanogenic massive sulfide deposits. Econ. Geol. **68**, 1321–1336 (1973)

Smewing, J.D., Simonian, K.O., Gass, I.G.: Metabasalts from the Troodos Massif, Cyprus: genetic implication deduced from petrography and trace element geochemistry. Contrib. Mineral. Petrol. **51**, 49–64 (1975)

Smith, A.B.: Alpine deformation and the oceanic areas of the Tethys, Mediterranean and Atlantic. Geol. Soc. Am. Bull. **82**, 2039–2070 (1971)

Smith, C.H.: Bay of Islands igneous complex, western Newfoundland. Geol. Survey Can. Mem. **290**, 1–132 (1958)

Smith, J.W., Green, D.H.: Geology of the Musa River area, Papua. Australia Bur. Mineral Resources, Geol. Geophys. Rept. **52**, 1961, 41 p.

Smith, P.J.: Ophiolites and oceanic lithosphere. Nature (London) **250**, 99–100 (1974)

Smith, P.J.: Disagreement over Troodos. Nature (London) **255**, 192–194 (1975)

Smith, R.E.: Redistribution of major elements in the alteration of some basic lavas during burial metamorphism. J. Petrol. **9**, 191–219 (1968)

Snavely, P.D., Jr., MacLeod, N.S., Wagner, H.C.: Tholeiitic and alkalic basalts of the Eocene Siletz River Volcanics, Oregon Coast Range. Am. J. Sci. **266**, 454–481 (1968)

Spooner, E.T.C.: Sub-sea-floor metamorphism, heat and mass transfer; an additional comment. Contr. Mineral. Petrol. **45**, 169–173 (1974)

Spooner,E.T.C., Beckinsale,R.D., Fyfe,W.S., Smewing,J.D.: O^{18}-enriched ophiolitic metabasic rocks from E. Liguria (Italy), Pindos (Greece), and Troodos (Cyprus). Contr. Mineral. Petrol. **47**, 41–62 (1974)

Spooner,E.T.C., Fyfe,W.S.: Sub-sea-floor metamorphism, heat and mass transfer. Contr. Mineral. Petrol. **42**, 287–304 (1973)

St.John,V.P.: The gravity field and structure of Papua and New Guinea. APEA J. **10**, 41–55 (1970)

Staub,R.: Über die Verteilung der Serpentine in den alpinen Ophioliten. Schweiz. Min. Petr. Mitt. **2**, 78–199 (1922)

Steinmann,G.: Geologische Beobachtungen in den Alpen (II). Die Schardtsche Überfaltungstheorie und die geologische Bedeutung der Tiefseeabsätze und der ophiolithischen Massengesteine. Ber. Natf. Ges. Freiburg i. B. **16**, 1–49 (1906)

Steinmann,G.: Die ophiolithischen Zonen in dem mediterranen Kettengebirge. 14th Intern. Geol. Congr. Madrid **2**, 638–667 (1927)

Stevens,R.K.: Cambro-Ordovician flysch sedimentation and tectonics in west Newfoundland and their possible bearing on a proto-Atlantic ocean. Geol. Assoc. Can. Spec. Paper **7**, 165–177 (1970)

Stocklin,J.: Possible ancient continental margins in Iran, In: The Geology of Continental Margins. Burk,C.A., Drake,C.L. (eds.). New York-Heidelberg-Berlin: Springer 1974, pp. 873–887

Stonely,R.: On the origin of ophiolite Complexes in the southern Tethys region. Tectonophysics **25**, 303–322 (1975)

Streckeisen,A.: Clasification and nomenclature of plutonic rocks. IUGS subcommission on systematic of igneous rocks. Geotimes **18** (10), 26–30 (1973)

Stueber,A.M.: Abundances of K, Rb, Sr and Sr isotopes in ultramafic rocks and minerals from western North Carolina. Geochim. Cosmochim. Acta **33**, 543–553 (1969)

Stueber,A.M., Murthy,V.R.: Strontium isotope and alkali element abundances in ultramafic rocks. Geochim. Cosmochim. Acta **30**, 1243–2159 (1966)

Suess,E.: Das Antlitz der Erde. Freytag, Leipzig **3** (2), 1–789 (1909)

Sugisaki,R., Mizutani,S., Adachi,M., Hattori,H., Tanaka,T.: Rifting in the Japanese Late Palaeozoic geosyncline. Nature Physical Sci. **233**, 30–31 (1971)

Sugisaki,R., Mizutani,S., Hattori,H., Adachi,M., Tanaka,T.: Late Paleozoic geosynclinal basalt and tectonism in the Japanese Islands. Tectonophysics **14**, 35–56 (1972)

Sutton,G.H., Maynard,G.L., Hussong,D.M.: Widespread occurrence of a high-velocity basal layer in the Pacific crust found with repetitive sources and sonobuoys. In: The Structure and Physical Properties of the Earth's Crust. AGU Geophys. Mono. Ser. **14**, 193–209 (1971)

Suzuki,T., Kashima,N., Hada,S., Umemura,H.: Geosyncline volcanism of the Mikabu green-rocks in the Okuki Area, Western Shikoku, Japan. Japanese Assoc. Mineralogists, Petrologists and Econ. Geol. J. **67**, 177–192 (1972)

Talwani,M., Le Pichon,X., Ewing,M.: Crustal structure of the midocean ridges 2. Computed model from gravity and seismic refraction data. J. Geophys. Res. **70**, 341–352 (1965)

Talwani,M., Windisch,C.C., Langseth,M.G.,Jr,: Reykjanes ridge crest: A detailed geophysical study. J. Geophys. Res. **76**, 473–517 (1971)

Taylor,H.P.,Jr.: The oxygen isotope geochemistry of igneous rocks. Contr. Mineral. Petrol. **19**, 1–71 (1968)

Taylor,H.P.,Jr.: Oxygen isotope evidence for large-scale interaction between meteoric ground waters and Tertiary granodiorite intrusions, western Cascade range, Oregon. J. Geophys. Res. **76**, 7855–7874 (1971)

Taylor, H. P., Jr., Epstein, S.: O^{18}/O^{16} ratios in rocks and coexisting minerals of the Skaergaard intrusion, E. Greenland. J. Petrol. **4**, 51–74 (1963)

Thayer, T. P.: Chromite deposits of Grant Co., Oregon, A preliminary report. U.S. Geol. Surv. Bull. **922**-D, 75–113 (1940)

Thayer, T. P.: Chrome resources of Cuba. U.S. Geol. Surv. Bull. **935**-A, 1–74 (1942)

Thayer, T. P.: Some critical differences between alpine-type and stratiform peridotite-gabbro complexes. Intern. Geol. Congr. 21st Sess. Copenhagen **13**, 247–259 (1960)

Thayer, T. P.: Flow-layering in alpine peridotite-gabbro complexes. Miner. Soc. Am. Spec. Paper **1**, 55–61 (1963)

Thayer, T. P.: Principal features and origin of podiform chromite deposits and some observations on the Guleman-Soridag district, Turkey. Econ. Geol. **59**, 1497–1524 (1964)

Thayer, T. P.: Chemical and structural relations of ultramafic and feldspathic rocks in alpine intrusive complexes. In: Ultramafic and Related Rocks. Wyllie, P. J. (ed.). New York: Wiley 1967, pp. 222–238

Thayer, T. P.: Gravity differentiation and magmatic re-emplacement of podiform chromite deposits. In: Wilson, H. D. B. (ed.). Magmatic Ore Deposits, Econ. Geol. Mon. **4**, 132–146 (1969a)

Thayer, T. P.: Peridotite-gabbro complexes as keys to petrology of mid-oceanic ridges. Geol. Soc. Am. Bull. **80**, 1515–1522 (1969b)

Thayer, T. P.: Chromite segregations as petrogenetic indicators. Geol. Soc. South Africa, Symposium on the Bushveld Igneous Complex and other Layered Intrusions, Spec. Pub. **1**, 380–390 (1970)

Thayer, T. P.: Gabbro and epidiorite versus granulite and amphibolite: a problem of the ophiolite assemblage. VI Caribbean Conf. Proc., Margarita, Venezuela, p. 315–320 (1972)

Thayer, T. P.: Some implications of sheeted dike swarms in alpine peridotite-gabbro complexes. Paper presented at International Symposium on Ophiolites in the Earth's Crust. Moscow: Academic of Sciences USSR 1974

Thayer, T. P., Brown, C. E.: Is the Tinaguillo, Venezuela, "Pseudogabbro" metamorphic or magmatic? Geol. Soc. Am. Bull. **72**, 1565–1570 (1961)

Thayer, T. P., Himmelberg, G. R.: Rock succession in the alpinetype mafic complex at Canyon Mountain, Oregon. 23rd Intern. Geol. Cong., **1**, 175–186 (1968)

Thompson, G. A.: Aeromagnetic and Bouguer gravity map of Twin Sisters dunite, Northwestern Washington. U.S. Geol. Surv. Geophys. Inves. Map GP-901, 1973

Thompson, G. A., Robinson, R.: Gravity and magnetic investigation of the Twin Sisters Dunite, Northern Washington. Geol. Soc. Am. Bull. **86**, 1413–1422 (1975)

Thompson, J. E.: A geological history of eastern New Guinea. APEA J. 83–93 (1967)

Thompson, J. E., Fisher, N. H.: Mineral deposits of New Guinea and Papua and their tectonic setting. Proc. 8th Comm. Min. Metall. Congr. 8th, Australia-New Zealand **6**, 115–148 (1965)

Thurston, D. R.: Studies on bedded cherts. Contr. Mineral. Petrol. **36**, 329–334 (1972)

Tobisch, O. T.: Gneissic amphibolite at Las Palmas, Puerto Rico, and its significance in the early history of The Greater Antilles Island Arc. Geol. Soc. Am. Bull. **79**, 557–574 (1968)

Trescases, J. J.: Premieres observations sur l'alteration des peridotites de Nouvelle-Caledonie. ORSTROM Ser. Geol. **1**, 27–58 (1969)

Trommsdorff, V., Evans, B. W.: Progressive metamorphism of antigorite schist in the Bergell Tonalite Aureole (Italy). Am. J. Sci. **272**, 423–437 (1972)

Trommsdorff, V., Evans, B. W.: Alpine metamorphism. of peridotitic rocks. Schweizer. Mineralog. Petrog. Mitt. **54**, 333–352 (1974)

Tschopp, R. H.: The general geology of Oman. Proc. 7th World Petrol. Congr., Mexico **2**, 231–242 (1967)

Turcotte, D. L., Oxburg, E. R.: Mantle convection and the new global tectonics. Ann. Rev. Fluid Mech. **4**, 33–68 (1972)

Upadhyay, H. D., Strong, D. F.: Geological setting of the Betts Cove copper deposits, Newfoundland: An example of ophiolite sulfide mineralization. Econ. Geol. **68**, 161–167 (1973)

Vallance, T. G.: Spilitic degradation of a tholeiitic basalt. J. Petrol. **15**, 79–96 (1974)

Viljoen, R. P., Viljoen, M. J.: The geology and geochemistry of the lower ultramafic unit of the Onverwacht Group and a proposed new class of igneous rocks in: Upper Mantle Project. Geol. Soc. S. Africa Spec. Publ. **2**, 55 (1969)

Vine, F. J., Matthews, D. H.: Magnetic anomalies over oceanic ridges. Nature (London) **199**, 947–949 (1963)

Vine, F. J., Poster, C. K., Gass, I. G.: Aeromagnetic survey of the Troodos igneous massif, Cyprus. Nature Phys. Sci. **244**, 34–38 (1973)

Vletter, R. de: How Cuban nickel ore was formed—a lesson in laterite genesis. Engin. Mining J. **156**, 84–87 (1955)

Vuagnat, M., Cogulu, E.: Quelques reflexions sur le massif basique-ultrabasique du Kizil Dagh, Hatay, Turquie. Soc. Phys. Hist. Natur. Geneve, C.R. **2**, 210–216 (1968)

Walcott, R. I.: Geology of the Red Hill Complex, Nelson, New Zealand. Royal Soc. New Zealand Trans. **7**, 57–88 (1969)

Walker, G. P. L.: Intrusive sheet swarms and the identity of crustal layer 3 in Iceland. J. Geol. Soc. London **131**, 143–161 (1975)

Weaver, D. F.: A geological interpretation of the Bouguer anomaly field of Newfoundland. Can. Dominion Obser. **35**, 223–251 (1967)

Wenner, D. B., Taylor, H. P., Jr.: δD and δO^{18} studies in serpentinization of ultramafic rocks (abstr.). Geol. Soc. Am. Atlantic City, 234–235 (1969)

Wenner, D. B., Taylor, H. P., Jr.: Temperatures of serpentinization of ultramafic rocks based on O^{18}/O^{16} fractionation between coexisting serpentine and magnetite. Contr. Mineral. Petrol. **32**, 165–185 (1971)

Wenner, D. B., Taylor, H. P., Jr.: Oxygen and hydrogen isotope studies of the serpentinization of ultramafic rocks in oceanic environments and continental ophiolite complexes. Am. J. Sci. **273**, 207–239 (1973)

Wenner, D. B., Taylor, H. P., Jr.: D/H and O^{18}/O^{16} studies of serpentinization of ultramafic-rocks. Geochim. Cosmochim. Acta, **38** 1255–1286 (1974)

Williams, H.: Mafic ultramafic complexes in western Newfoundland Appalachians and the evidence for their transportation: A review and interim report. Geol. Assoc. Can. Proc. (A Newfoundland decade) **24**, 9–25 (1971)

Williams, H.: Bay of Islands, map-area, Newfoundland. Geol. Surv. Canada, Paper **72–34**, 1973, 7 p.

Williams, H., Malpas, J.: Sheeted dikes and brecciated dike rocks within transported igneous complexes, Bay of Islands, Western Newfoundland. Can. J. Earth Sci. **9**, 1216–1229 (1972)

Williams, H., Smyth, W. R.: Metamorphic aureoles beneath ophiolite suites and Alpine peridotites: Tectonic implications with west Newfoundland examples. Am. J. Sci. **273**, 594–621 (1973)

Williams, H., Stevens, R. K.: The ancient continental margin of eastern North America. In: The Geology of Continental Margins. Burk, C. A., Drake, C. L. (eds.). Berlin-Heidelberg-New York: Springer 1974, pp. 781–796

Wilson, H. H.: Late Cretaceous eugeosynclinal sedimentation, gravity tectonics, and ophiolite emplacement in Oman Mountains, Southeast Arabia. Am. Assoc. Petrol. Geol. Bull. **53**, 626–671 (1969)

Wilson, R. A. M.: The geology of the Xeros-Troodos area. Cyprus Geol. Survey Dept. Mem. **1**, 1959, 184 p.

Winkler, H. G. F.: Petrogenesis of Metamorphic Rocks. New York-Heidelberg-Berlin: Springer, 3rd ed. (1974)

Wyllie, P. J. (ed.): Ultramafic and Related Rocks. New York: Wiley 1967, 464 p.

Yoder, H. S., Jr.: Spilites and serpentinites. Carnegie Inst. Washington Year Book **65**, 269–279 (1967)

Zimmerman, J., Jr.: Emplacement of the Vourinos ophiolitic complex, northern Greece. Geol. Soc. Am. Mem. **132**, 225–239 (1972)

Subject Index

Springer-Verlag
Berlin
Heidelberg
New York

PHYSICS AND CHEMISTRY OF MINERALS

A new journal

Edited by S.S. Hafner, A.S. Marfunin, C.T. Prewitt

Physics & Chemistry of Minerals

is an international journal devoted to publishing articles and short communications
of physical or chemical studies on minerals or solids related to minerals. The
aim of the journal is to support competent interdisciplinary work in mineralogy
and physics or chemistry. Particular emphasis is placed on applications of modern
techniques or new theories and models to interpret atomic structures and physical
or chemical properties of minerals. Some subjects of interest are:
- Relationships between atomic structure and crystalline state (structure of various
  states, crystal energies, crystal growth, thermodynamic studies, phase
  transformations, solid solution, exsolution phenomena, etc.)
- General solid state spectroscopy (ultraviolet, visible, infrared, Raman, ESCA,
  luminescence, X-ray, electron paramagnetic resonance, nuclear magnetic
  resonance, gamma ray resonance, etc.)
- Experimental and theoretical estimation of chemical bonds in minerals
  (application of crystal field, molecular orbital, band theories, etc.)
- Physical properties (mechanic, magnetic, electric, optical, etc.)
- Relations between thermal expansion, compressibility, elastic constants, and
  fundamental properties of atomic structure, particularly as applied to geophysical
  problems.
- Electron microscopy in support of physical and chemical studies.

Springer-Verlag Berlin Heidelberg New York

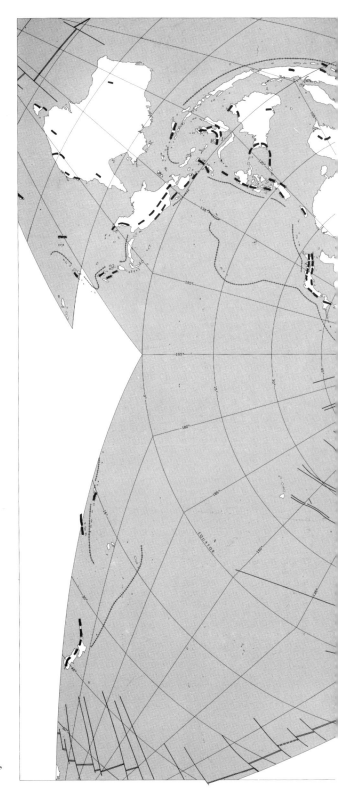

North polar projection showing
the principal ophiolite belts
of the world.
By W. P. Irwin and R. G. Coleman,
U.S. Geological Survey, 1974.